SELECTED EXERCISES IN GALACTIC ASTRONOMY

ASTROPHYSICS AND SPACE SCIENCE LIBRARY

VOLUME 26

I. ATANASIJEVIĆ

SELECTED EXERCISES
IN
GALACTIC ASTRONOMY

D. REIDEL PUBLISHING COMPANY

DORDRECHT-HOLLAND

Library of Congress Catalog Card Number 73–159652

ISBN-13: 978-94-010-3113-4   e-ISBN-13: 978-94-010-3111-0
DOI: 10.1007/978-94-010-3111-0

TABLE OF CONTENTS

INTRODUCTION

Galactic Astronomy is, and for some time to come will certainly remain, one of the more important items in the astronomy curricula. There are at present a number of excellent text-books and monographs covering the whole subject or certain of its aspects. It seems however that there is not a single book where at least some of the more important problems dealt with in text-books and university courses would be presented in the form of laboratory exercises. This short series of exercises represents an attempt to fill this gap.

What is, in general, the aim of such exercises? As to this point, the author fully agrees with Prof. Minnaert's opinion that "No natural science should ever be taught without practical work. Students in astronomy should have regular exercises, not so much to teach them observing skill, but mainly to bring before their eyes the reality of the concepts introduced during the lectures." This applies to Galactic Astronomy as well as to any other branch of our science.

In particular, as far as Galactic Astronomy is concerned, the author is deeply convinced that the student must be given the opportunity to see some of the phenomena hidden behind the columns of numbers filling the astronomical catalogues and the tables in many publications, and which for an inexperienced eye look rather dull and uninteresting. He also believes that it is useful and necessary to invite the student to take an active part in the exploration of the subject. Or, to be more specific, the student shall not only be told how, in principle, Prof. A or Dr. B have determined this or that quantity or discovered this or that effect. He shall, moreover, be asked to take the astronomical catalogues or the papers where the relevant primary data have been published or can be found, and to compute, like his famous predecessors, some specific quantity or to verify the existence of some effect. In general it can be said that each of the exercises initiates the student to the essence of an original investigation and at the same time expounds some of the fundamental techniques used in Stellar Astronomy. The author believes that such guided exercises may serve as a useful preparation for independent research work.

The author regrets that the present book does not cover all the questions dealt with in a standard course in Galactic Astronomy. Perhaps this is even not desirable as within the framework of such a course a large part must be left for various forms of individual independent work on the part of the student. Problems of (apparent) distribution of the stars are dealt with in a rather elementary manner in the first two exercises, and interstellar absorption is only briefly considered. The main accent is on the questions pertaining to stellar motions, up to the computation of the galactic orbit of a star. No direct use has been made of the results of Radio Astronomy, but they have been considered in the discussion. Some interesting and important questions have been omitted because the primary data were not generally available in an appropriate form, other ones in view of the prohibitively large amount of work involved. It is hoped that even so this book may contribute to a better understanding of the problems and methods of Galactic Astronomy.

As to the form in which the exercises have been presented, the author would like to recall that there is, in general, a rather wide gap between the account of a method which, within the framework of a lecture (or of a text-book), can be considered as fully acceptable, and the procedure which, in practice, one effectively has to follow in order to arrive at a useful result. It is obvious that in the last case one is compelled to discuss many technical details which in a text-book can safely be omitted. The author does not believe that it would serve the purpose to refer the reader, for the derivation of the fundamental formulae which he will have to use, to various sources. In fact, some of them may not be available, other ones may prove to be too hard to understand, as they have not been written for the student; the notations and the conventions adopted may be different. Therefore all the formulae have been derived from first principles. However this does not at all imply that the reader shall not carefully consult the monographs or the papers cited. Quite on the contrary, the reader is expressly advised to do so. The introductions (and quotations) are intended to pave the road towards the understanding of the original papers. It cannot be denied that, such as they are, the introductions may seem rather long. However the author must confess that he prefers articles and books where the assumptions made are clearly stated and discussed, and all the equations derived, to those qualifying too many statements as well known or obvious, and most of the formulae used as easy to prove or proved elsewhere.

In view of the fact that the treatment of numerical data is, in general, a somewhat neglected subject, some of the questions pertaining to it have also been included for the reader's convenience. The general background expected is that of a student following more advanced courses in astronomy. Obvious prerequisites are general courses in astronomy, mathematics and mechanics.

The primary data to be used have been (or shall be) taken from astronomical catalogues or original publications. No attempt has been made to select amongst them only the best ones, i.e., those which will give the best fit, as this is considered as an inadmissible forgery. Whenever this has been possible tables containing the primary data have been given. This, too, has been done for the reader's convenience, as it is not expected that more than one copy of the original papers or catalogues will be available. The reader is again warned that this does not mean that the originals should not be consulted.

The choice of the references has been determined by didactic considerations. They do not represent a complete and up-to-date bibliography on the subject discussed, but rather a choice which, in the author's opinion, meets the needs and the capabilities of the reader.

All exercises except the last one can be carried out using any standard desk calculator. For the numerical integration in Exercise VIII one really needs a programmable electronic desk computer. If possible all exercises involving longer calculations shall be made with the aid of such a desk computer making extensive use of its programming capabilities.

Whenever graphical methods of solution are proposed, the kind of graph which the reader should obtain is explained by a schematical sketch. The numerical solutions are given in detail. Of course they should only be consulted to check one's own solution or to find computation errors.

The problems added at the end of some of the exercises are mainly intended to complete or to clarify some of the points treated in the respective exercises. A very complete collection of 59 more advanced problems (with solutions) can be found in the third edition of Prof. P. P. Parenago's excellent *Textbook of Stellar Astronomy* (in Russian, Moscow, 1954).

Most of the exercises cannot be carried out in the time usually available for scheduled laboratory exercises. The student is advised first to read carefully the whole text and then to prepare all he will need for his work (tables, catalogues, graph paper, the desk calculator with instruction for use, and so on). Somewhere in the astronomy department he must be given a quiet place where he can work for some hours during several days.

With the enthusiastic collaboration of Mrs. J. Milogradov-Turin, assistant at the Faculty of Sciences, Belgrade, a first series of exercises has been prepared, tested in effective work with the students of that Faculty, and reported at the General Discussion on the Teaching of Astronomy, at the XII General Assembly of the IAU (*Trans. IAU* **XIIB** (1964), p. 641). The present book is a thoroughly revised, enlarged and modified version of this first series.

The author is deeply indebted to the late Prof. M. G. J. Minnaert, Utrecht, for encouraging this work, the keen interest he has shown, and for many fruitful discussions.

Prof. L. Plaut, Groningen, has read the larger part of the manuscript, Dr. A. Ollongren, Leiden, the last exercise. To both of them the author expresses his sincere thanks for their help and the valuable suggestions they have made.

All the exercises have effectively been used in a course given by the author during several years at the Faculty of Sciences, of the Catholic University, Nijmegen, The Netherlands.

GENERAL REFERENCES

The most recent text book which covers the whole field of Galactic Astronomy is
  Mihalas, D., with the collaboration of McRae Routly, P.: 1968, *Galactic Astronomy*, Freeman and Co., San Francisco and London.

Classical problems in Stellar kinematics are extensively discussed in
  Smart, W. M.: 1968, *Stellar Kinematics*, Longmans, Green and Co., London.

Parenago's text book has appeared in three editions the last of which is
  Parenago, P. P.: 1954, *A Course in Stellar Astronomy* (in Russian), GTTI, Moscow.

Another very useful text book in Russian (translated from Polish) is
  Zonn, W. and Rudnicki, K.: 1959, *Stellar Astronomy*, IL, Moscow.

Although first published in 1953, the monograph
  Trumpler, J. and Weaver, H. F.: 1962, *Statistical Astronomy*, Dover Publ. Inc., New York.
remains in many respects a classic. However this is not a book for the beginner.

A more recent monograph on the same subject, especially interesting for a mathematically minded reader is
  Kurth, R.: 1967, *Introduction to Stellar Statistics*, Pergamon Press, Oxford.

The introductory chapters of
  Ogorodnikov, K. F.: 1965, *Dynamics of Stellar Systems*, Pergamon Press, Oxford
contain a very clear and complete account of the principal problems of stellar kinematics as well as of the basic ideas of stellar statistics.

Volume V of the series Stars and Stellar Systems,
  Blaauw, A. and Schmidt, M. (eds.): 1965, *Galactic Structure*, University of Chicago Press, Chicago
must be considered as the most authoritative source of informations on the results and problems in galactic research in the early sixtieth.

The most important numerical data can be found in
  Allen, C. W.: 1963, *Astrophysical Quantities*, 2nd ed., The Athlone Press, London.

For more detailed information the reader should consult:
  Voigt, H. H. (ed.): 1965, Landolt-Börnstein, Numerical Data, new series, group VI/1, *Astronomy*, Springer-Verlag, Berlin, Heidelberg and New York.

The three principal abstracting journals in astronomy prepared under the auspices of the International Astronomical Union, are
  Astronomy and Astrophysics Abstracts, produced by the Astronomisches Rechen-Institut, Heidelberg, Springer-Verlag, Berlin, appearing in semi-annual volumes;
  Bulletin Signalétique: Astronomie, published by the Centre de Documentation du C.N.R.S., Paris, ten issues per year;
  Referativnyj Zhurnal: Astronomiya, published by the Institute for Scientific and Technical Information of the U.S.S.R., Moscow, twelve issues per year.

For publications which have appeared between 1898 and 1969, the reader may consult
   Astronomischer Jahresbericht, ed. by the Astronomisches Rechen-Institut, Heidelberg, and published by W. de Gruyter, Berlin, annually, up to 1969.

The advances in all fields of astronomical research are briefly reviewed in the reports which the presidents of the respective commissions of the International Astronomical Union prepare for each General Assembly of the Union. They are published, with full bibliographical references, in the volumes of the

Transactions of the International Astronomical Union

after each General Assembly (present publisher: Reidel, Dordrecht, The Netherlands).

EXERCISE I

DETERMINATION OF THE POSITION OF
THE GALACTIC EQUATOR

1. In this first exercise we shall determine the position of the plane of symmetry of our Galaxy. Furthermore we shall define a system of spherical coordinates especially appropriate for the study of the distribution and motions of objects belonging to our stellar system. For obvious reasons the plane of symmetry will be adopted as the fundamental plane of this coordinate system.

Let us first briefly recall some of the most important observational facts on which our determination will be based. There is, first of all, the belt of the Milky Way, certainly the most conspicuous large-scale phenomenon on the sky. Turning to more refined observations remember that, as shown by the star counts, the number per square degree of all stars down to a certain magnitude varies over the sky in a characteristic way. This number rises steeply when approaching the Milky Way where it attains a maximum. Instead of such a wide class of objects as all the stars brighter than a certain limiting magnitude, one can choose a homogeneous group of objects and then investigate their distribution over the sky. For instance one can take the galactic clusters, or the variables belonging to a well defined type or the stars of a definite spectral type and luminosity class. In general the degree of concentration towards the Milky Way will differ from one group to the other, objects of certain classes being confined to the Milky Way, whereas others are scattered widely over the sky. In any case, however, it is found that the crowding towards the Milky Way is a feature common to all of them which, therefore, must correspond to an essential characteristic of our stellar system.

Non-optical investigations corroborate this evidence. On all isophotal maps obtained by radio-astronomical surveys of the continuum radiation one easily recognises a ridge line corresponding to the belt of the Milky Way. At the opposite end of the spectrum, most of the recently discovered cosmic X-ray sources exhibit a strong concentration towards the Milky Way.

In order to go beyond a mere verbal description of facts one shall, first, try to express them in a quantitative, mathematical form, even at the cost of some idealisation; and, second, to propose an interpretation. In this respect it is important that a great circle can be drawn to which the course of the Milky Way most nearly conforms. Moreover all objects belonging to our Galaxy are symmetrically distributed with respect to this circle which is called the galactic equator.

Remembering that to any great circle on the sphere there corresponds a plane drawn through the position of the observer, such a distribution can be interpreted as follows:

(a) All objects belong to our Galaxy are distributed symmetrically with respect to a plane, which is the plane of symmetry of our stellar system; and,

(b) The position we occupy is so close to the plane of symmetry that, at least in a first approximation, our distance from it can be neglected.

2. In mathematical terms our problem can be stated as follows:

Given the distribution of a class of objects over the celestial sphere, determine the position of a great circle best fitting this distribution.

The problem being stated, we have to consider more closely the following two questions:

First, which group of objects shall we choose, and,

Second how a 'best-fitting circle' shall be defined and determined.

What the first question is concerned, we obviously shall expect to get the best results choosing objects which exhibit the strongest concentration towards the Milky Way. It is, however, clear that, in a more general investigation, other objects too must be considered. Our determination will be based on the study of the apparent distribution of the young Population I objects, and particularly the youngest among them. We propose the galactic clusters, the Wolf-Rayet and O-type stars. The classical Cepheids can also be used.

We shall need lists of these objects containing their right ascensions and declinations.

Now it is important to realise that the position of the galactic equator as deduced from the distribution of any group of objects will refer to the same frame of reference as the coordinates of these objects. In other words if, for instance, we have used a list where the coordinates are given for the epoch 1950.0, then we shall get the position of the galactic equator with respect to the celestial equator and equinox for the epoch 1950.0. As due to precession this frame is moving with respect to the whole of the stars, any determination based on coordinates for a different epoch will necessarily lead to a different result. Therefore, before comparing the results deduced for different epochs we must take into account the effect of precessional motion.

Let us now quote some sources.

A useful list of galactic clusters can be found in A. Bečvář's *Atlas Coeli II – Katalog 1950.0* [1]. It has 293 entries. The coordinates are given for the epoch 1950.0.

A more recent and complete catalogue of galactic clusters has been compiled by H. Sawyer Hogg [2]. This catalogue contains 514 clusters, most of them classified according to a scheme proposed by R. Trumpler (ibidem, p. 135). As the class I clusters are among the youngest we can select them for a determination of the position of the galactic equator. The equatorial coordinates are given for the epoch 1950.0.

The most complete catalogue of star clusters and associations is that compiled (and regularly enlarged) by G. Alter *et al.* [3].

Recent lists of Wolf-Rayet stars have been given by M. C. Roberts [4], as well as by L. F. Smith [5]. Both give the coordinates for the epoch 1900.0.

In order to make a complete list of the classical Cepheids (or δ Cephei variables) we should have to start from the list of all cepheids given in the second volume of the *General Catalogue of Variable Stars* [6]. Among them we should have to choose these belonging to the young Population I (see Preface to Vol. II of the General Catalogue), and then look for their coordinates in Vol. I. In that volume the variables are grouped according to the alphabetical order of the constellation names. The epoch is 1900.0. The General Catalogue represents the standard reference book for investigations on variable stars. A new edition has just been published.

A shorter list of δ Cephei variables based on data given by R. P. Kraft and M. Schmidt in an important paper [7], is given in Table I.1. (epoch 1900.0) (see pp. 17–19).

3. The next question is how to define and to determine the great circle best fitting the distribution of the objects chosen over the sphere. We recall that the position of any great circle is given either by the coordinates of its poles, or, if the circle is oriented, by the right ascension of its ascending node and its inclination to the celestial equator. Our task will therefore be to deduce, from the apparent distribution of our objects, one of the two pairs of quantities.

This problem can be solved either by numerical or by graphical methods. If a numerical method is adopted, one has first to specify, in mathematical terms, the exact meaning of the requirement that the fit shall be the best one. When this has been done, it is not difficult to deduce the equations leading to the solution. Due to the rather prohibitive amount of computational work needed, we will not use numerical methods. Nevertheless, and in view of the importance of the problem, as well as of the fact that graphical methods are somewhat subjective, we propose that the reader should derive the principal equations of the numerical method due to S. Newcomb (See Problem No. 1 at the end of this exercise).

We shall adopt a simple and straightforward graphical method giving quite good results.

On two sheets of transparent graph paper represent all the objects chosen, taking as abscissae their right ascensions and as ordinates their declinations. The x-axis will therefore correspond to the celestial equator. One sheet shall contain all objects between 0 h and 14 h, the other one those between 12 h and 2 h. In other words we shall map our objects in a somewhat unusual, but for this purpose quite convenient, cartographic projection. As we will choose objects exhibiting a very pronounced galactic concentration, the representative points will occupy a rather narrow band.

On both sheets trace, by free hand, a line best fitting the representative points. How this should be done depends on your judgement. It is this that makes the whole

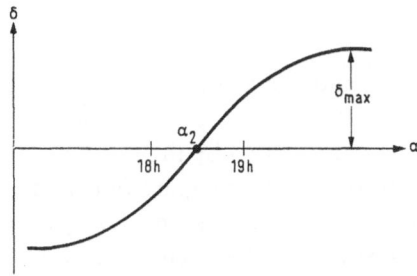

Fig. I.1.

determination somewhat subjective. The fairly symmetric curve will be found to have a maximum between 0 h and 1 h, and a minimum between 12 h and 13 h, reaching very approximatively the same distances from the celestial equator. It will intersect the celestial equator in two very nearly opposite points, one between 6 h and 7 h, the other one between 18 h and 19 h.

Let α_1 be the right ascension of the point of intersection between 6 h and 7 h, and α_2 the right ascension of the other one. For the right ascension of the point of intersection lying between 18 h and 19 h (in the constellation of Aquila) we will adopt the value

$$\alpha_{gn} = \tfrac{1}{2}(\alpha_1 + 12 \text{ h} + \alpha_2). \tag{I.1}$$

By convention [8] this point has been taken as the ascending node of the galactic equator on the celestial equator. It is important to realise that by this convention we have at the same time implicitly oriented the galactic equator, that is defined a positive direction along it. In fact, according to its definition, the ascending node (on the celestial equator) is the point where, moving in the positive direction along the galactic equator, we cross the celestial equator going north.

Let us now deduce the inclination of the galactic equator. Obviously the inclination is equal to the declination of the maximum. As above, and due to the symmetry of the curve of best fit, we shall expect to get a better value for the inclination i taking

$$i = \tfrac{1}{2}(\delta_{\max} + |\delta_{\min}|). \tag{I.2}$$

These two elements which define the position of the galactic equator with respect to the celestial equator can also be found as follows.

Place the two sheets face to face. Turn one of them top to bottom. Superpose as exactly as possible the two segments of the x-axis, taking care that points corresponding to right ascension α h come exactly under those with right ascension $\alpha + 12$ h. By this device we have obtained that points diametrically opposite on the celestial sphere are exactly superposed.

Now observe the two narrow bands of representative points and examine how far they, too, are superposed. As you will see this will be the case to a very good degree of approximation. This being so we can trace now on the back side of one sheet (looking through both) the curve of best fit and deduce the right ascension of the point of intersection as well as the declination of the maximum.

Note the very patchy distribution of the representative points along the galactic equator.

4. Although the values of α_{gn} and i have been determined, we cannot yet consider that our task has been completed. Before drawing any conclusion we must see how far the apparent distribution can really be fitted by the great circle which these values determine. In other words we now have to compare our model with reality. In order to achieve this we now shall trace on our diagrams this great circle.

Denote by α_{ge} and δ_{ge} respectively the right ascension and the declination of an arbitrary point lying on the provisionally adopted galactic equator. By elementary

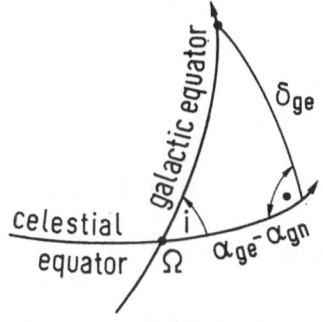

Fig. I.2.

spherical trigonometry one can deduce (Figure I.2.) that

$$\text{tg}\,\delta_{ge} = \text{tg}\,i\,\sin\left(\alpha_{ge} - \alpha_{gn}\right). \tag{I.3}$$

We shall now use this formula to calculate the values of δ_{ge} which correspond to $\alpha_{ge} - \alpha_{gn}$ equal to 0 h, 1 h, 2 h, ..., 6 h. From this it is easy to find the values of δ_{ge} for the remaining three quadrants. As always in this kind of calculations make use of an appropriate computing form. Here we propose the following one:

Equatorial coordinates of points on the provisional galactic equator

$$\text{tg}\,\delta_{ge} = \text{tg}\,i\cdot\sin\left(\alpha_{ge} - \alpha_{gn}\right)$$

$\alpha_{gn} =$	$i =$	$\text{tg}\,i =$			
$\alpha_{ge} - \alpha_{gn}$	$\sin\left(\alpha_{ge} - \alpha_{gn}\right)$		$\text{tg}\,\delta_{ge}$	α_{ge}	δ_{ge}
0 h					
1 h					
2 h					
..					
6 h					

The form has a heading indicating the calculations for which it is intended. Then comes the formula which shall be used. The next row contains the adopted values of the constants. The computing form proper has to be completed column by column, either by filling in values taken from tables of trigonometric functions (column II) or by computing the corresponding values by the formula given above (column III). The values in column IV (α_{ge}) will obviously be $\alpha_{gn} + 0\,\text{h} = \alpha_{gn}$; $\alpha_{gn} + 1\,\text{h}$; ...; $\alpha_{gn} + 6\,\text{h}$. After having computed the seven values of tg δ_{ge} fill in the last column using tables of trigonometric functions.

One could perhaps say that a computing form together with the instruction of how it shall be used represents a primitive form of a computing programme.

With all the necessary values of α_{ge} and δ_{ge} trace, on the diagrams, the line corresponding to the great circles whose elements have been deduced, and appreciate the fit. It is possible, but not probable (if the work has been carefully done), that this curve deviates systematically from the apparent distribution. Should this be the case the whole determination must be repeated in order to get a better fit. On the other hand local deviation can be ascribed to special phenomena which in our very schematical model have not been taken into account.

Once the satisfactory fit has been obtained try to estimate the errors of your results. What is the range of the values of α_{gn} and of δ_{max} which could be allowed?

Deduce, furthermore, the position of the galactic pole which is in the north celestial hemisphere. Let α_{gp} and δ_{gp} respectively be the right ascension and the declination of this pole. Then it is easily seen that

$$\alpha_{gp} = \alpha_{gn} - 6\,\text{h}, \qquad \delta_{gp} = 90° - i. \tag{I.4}$$

This pole is called the North Galactic Pole.

Compare your results with the results of other determinations as well as with the standard values quoted in handbooks and textbooks. Do not forget to reduce them, if necessary, to the same epoch. This reduction is quite simple when the position of the galactic equator is given by the coordinates of its pole, because you can use any precessional tables (e.g. those given at the end of Bečvář's Katalog already cited).

State your conclusions concerning the possibility to interpret the apparent distribution by the simple model defined by the two assumptions stated under (a) and (b) at the end of Section 1.

5. Phenomena are more easily described and interpreted if adequate coordinate systems are used. It should, however, be stressed that a coordinate system is much more than a useful mathematical tool. It is well known, from Positional Astronomy, that the fundamental planes and directions, to which the fundamental great circles and zero points of the different coordinate systems are related, always have a deep physical meaning.

The coordinate system especially adapted to the needs of galactic research is the galactic coordinate system. We now shall consider some questions connected with its definition.

Let us first briefly recall the fairly simple procedure generally adopted when defining a system of spherical coordinates. Usually this is done as follows. One chooses a great circle as the fundamental circle of the coordinate system. Then on this circle a zero point as well as a positive direction are adopted. One of the two spherical coordinates is measured along the fundamental great circle, from the zero point and in the positive direction. The second coordinate is simply the spherical distance from the great circle or from its pole. Obviously, in order to avoid ambiguity, one must adopt a convention concerning the sign of the second coordinate.

Applying this scheme we will obviously adopt the galactic equator as the fundamental great circle of the galactic coordinate system. Now different determinations of its position with respect to the celestial equator lead to somewhat different results. Therefore the final choice of the great circle which has to be adopted as the fundamental one can only be made after a careful and critical examination of all available determination.

This problem has a long history. "William Herschel in 1785 first recognized the importance for investigations of the Milky Way of a special coordinate system having its equator in the galactic plane, and from this time until 1932 various systems having poles differing from each other by one or two degrees were used. The literature up to 1932 is summarized in the introduction to the well known Lund Observatory Conversion Tables [12] prepared by Ohlsson"* to which the reader is

* Cited from: 'The New System of IAU Galactic Coordinates', *Monthly Notices Roy. Astron. Soc.* **121**, 123.

referred for interesting historical details. The growing interest in galactic research made it imperative, in the thirties, to reach agreement about the choice of a standard system of coordinates.

At the Fifth General Assembly of the International Astronomical Union (Paris, 1935), the following recommendation has been adopted (the original is in French):

"That the Union should recommend the suggestion put forward by Dr. Parvulesco to adopt as the fundamental galactic pole the pole

$$\alpha = 12 \text{ h } 40 \text{ m} \qquad \delta = +28° \text{ (1900.0)}$$

which is used in Ohlsson's Tables of Galactic Coordinates" [9].

Now in his Conversion Tables Ohlsson had adopted the node in the constellation of Aquila as the ascending node of the galactic equator on the celestial equator, and, at the same time, as the zero point.

The coordinate counted along the galactic equator, from the zero point, and in the positive direction, is called the galactic longitude, and usually denoted by l. The spherical distance from the galactic equator, counted positive towards the north ('fundamental') pole is called the galactic latitude, and usually denoted by b.

The galactic coordinates cannot be directly measured, but instead they are computed from equatorial coordinates. It is not difficult to derive the necessary transformation formulae (see Problem No. 2). Approximate values can be obtained from nomograms, better ones from conversion tables, the most extensive being those by Ohlsson. Short excerpts from Ohlsson's Tables can be found in many textbooks and handbooks (e.g. [10]).

This coordinate system, extensively used up to 1958, is now called the old system. From 1958 on the coordinates which refer to it are designated by l^I and b^I. The new system adopted in 1958 will be defined in the next exercise. When using galactic coordinates or conversion tables make sure to which system they refer.

6. Let us finally discuss some weak points of our determination as well as the possible improvements.

The choice of the zero point from which galactic longitudes are counted is surely open to criticism. In fact this point does not correspond to any privileged direction in the galactic plane and has no physical meaning. On the other hand it is a fact that our previous investigations, limited to Population 1 objects, did not give any clear indication as to the existence of such a direction. The search for a physically meaningful zero point is therefore a separate task. In any case it will be greatly facilitated by the use of even a provisional system of galactic coordinates.

Another important point is the following one. Our determination has been based on the two assumptions formulated at the end of Section 1. At the degree of approximation adopted the results deduced do not contradict the facts. Now closer investigations show that, nevertheless, they represent only a first, although very good approximation. Therefore a second and better one can and shall be considered. In making this further step, one retains the first assumption concerning the space distribution of the various groups of objects. As to the second, the Sun's distance from the plane of symmetry is no longer considered as negligible. Instead it is taken as a quantity small when compared to the distance, from the Sun, of the objects investigated.

It is not difficult to see that in such a case the circle of best fit will be a small circle very close to a great one. Now a small circle cannot be adopted as the fundamental circle in a system of spherical coordinates. The galactic equator is then defined as the great circle parallel to it.

Now due to the one-to-one correspondence between circles on the sphere and planes cutting the sphere this problem (and similar ones) can be treated either by space geometry or by spherical trigonometry. Stating the problem in terms of space geometry one considers first the perpendicular drawn from the Sun through the plane of symmetry. The points where this perpendicular intersects the celestial sphere (concentric to the Sun) are the galactic poles. Then the galactic equator is defined as the great circle corresponding to these poles. For more details as well as for the derivation of the fundamental formula leading to the solution of this problem, the reader is referred to Problem No. 3 at the end of this Exercise.

Finally we propose that the more interested reader should try to find for himself the existence of another great circle, inclined to the galactic equator, towards which certain classes of close objects tend to crowd (Gould's Belt). Some indication of it can be found by investigating by the same simple graphic method the apparent distribution of diffuse nebulae listed in Bečvář's Katalog, or of B stars brighter than apparent magnitude 5. As a starting reference the reader can take [11] plunging for more recent data into the volumes of the standard astronomical abstracting journals (see: General References, p. XI) and current publications.

Problem I.1. Determination of Position of the Galactic Pole: Newcomb's Method

In his study 'On the Position of the Galactic and other Principal Planes Towards which the Stars Tend to Crowd' (Washington, D.C., 1904), S. Newcomb has thoroughly discussed the problem of the determination of the position of the galactic equator and derived the coordinates of its pole which are very close to the values adopted at present. Assuming that all stars are at the same distance from the Sun (which distance is taken as unity), S. Newcomb defines the plane of symmetry as follows:

Let us suppose a plane taken at pleasure passing through our position in the universe, which point we take as the origin of coordinates. This plane will cut the celestial sphere in a great circle. The perpendicular distance of a star will then be represented by the sine of its distance from the great circle. Let us form the sum of the squares of these sines for the whole system of stars which we consider. The value of this sum will vary with the position which we assign to the plane. The principal plane of condensation, as I define it, is that for which the sum in question is a minimum.

State the problem in terms of analytic geometry in space, considering the distances of the stars from the plane (all stars being at unit distance from the Sun). Derive the expression giving the sum of the squared distances and show that this sum is effectively a function of the position of the plane. Show that the problem is one of extreme values with side conditions and verify that the necessary conditions are

$$
\begin{aligned}
(a_{11} - \lambda)\, l + a_{12} m \quad\;\; + a_{13} n \quad\;\;\; &= 0 \\
a_{12} l \quad\;\; + (a_{22} - \lambda)\, m + a_{23} n \quad\;\; &= 0 \\
a_{13} l \quad\;\; + a_{23} m \quad\;\; + (a_{33} - \lambda)\, n &= 0
\end{aligned}
$$

where l, m, n, are the direction cosines of the perpendicular to the plane and where

$$
a_{11} = \sum_i x_i^2 \qquad a_{12} = \sum_i x_i y_i \qquad a_{13} = \sum_i x_i z_i
$$

$$
a_{22} = \sum_i y_i^2 \qquad a_{23} = \sum_i y_i z_i
$$

$$
a_{33} = \sum_i z_i^2
$$

are the sums of the squares and of the products of the coordinates of the stars. Show that λ is Lagrange's multiplier which can be computed from the cubic:

$$
\begin{vmatrix}
a_{11} - \lambda & a_{12} & a_{13} \\
a_{12} & a_{22} - \lambda & a_{23} \\
a_{13} & a_{23} & a_{33} - \lambda
\end{vmatrix} = 0.
$$

Solution

Take a coordinate system having its origin in the Sun and oriented as follows:

x-axis towards the point $\alpha = 0$ h, $\delta = 0°$
y-axis towards the point $\alpha = 6$ h, $\delta = 0°$
z-axis towards the North Celestial Pole

for a given epoch.

The normal equation of a plane is

$$lx + my + nz - p = 0, \tag{P.I.1}$$

where p is the distance of the plane from the origin, and where l, m, and n, are the direction cosines of the perpendicular to the plane drawn from the origin. They fulfil the equation:

$$l^2 + m^2 + n^2 = 1. \tag{P.I.2}$$

The equatorial (spherical) coordinates of the point towards which the perpendicular is directed, α_p, δ_p, are given by

$$\cos\delta_p \cos\alpha_p = l, \qquad \cos\delta_p \sin\alpha_p = m, \qquad \sin\delta_p = n. \tag{P.I.3}$$

The distance D_i of any point $P_i(x_i, y_i, z_i)$ from the plane is given by

$$D_i = l \cdot x_i + m \cdot y_i + n \cdot z_i - p. \tag{P.I.4}$$

Now let α_i, δ_i, be the equatorial coordinates of one of the stars investigated. Assuming that they all are at unit distance from the Sun we have

$$x_i = \cos\delta_i \cos\alpha_i, \qquad y_i = \cos\delta_i \sin\alpha_i, \qquad z_i = \sin\delta_i. \tag{P.I.5}$$

The distance of any star i from a plane drawn through the origin $(p=0)$ is

$$D_i = l \cdot x_i + m \cdot y_i + n \cdot z_i. \tag{P.I.6}$$

The sum of the squared distances from the plane is given by

$$\sum_i D_i^2 = \sum_i (l \cdot x_i + m \cdot y_i + n \cdot z_i)^2 = F(l, m, n), \tag{P.I.7}$$

which obviously is a function of the direction cosines of the perpendicular, l, m, and n, i.e. of the orientation of the plane.

The problem is to find a set of values of l, m, and n, which will make this expression a minimum. Now we cannot simply equate to zero the partial derivatives of (P.I.7) with respect to l, m, and n, because these variables are subject to the side condition (P.I.2) which we will write

$$f(l, m, n) = l^2 + n^2 + n^2 - 1 = 0. \tag{P.I.8}$$

The problem is thus a typical one of extreme values with side conditions. The solution is most easily found by Lagrange's method of indeterminate multiplyers. For that purpose multiply (P.I.8) with $-\lambda$ and add to (P.I.7). This gives

$$\bar{F}(l, m, n) = F(l, m, n) - \lambda f(l, m, n) \qquad \text{(P.I.9)}$$

or explicitly

$$\bar{F}(l, m, n) = (a_{11} - \lambda) l^2 + (a_{22} - \lambda) m^2 + (a_{33} - \lambda) n^2 \\ + 2(a_{12}lm + a_{13}ln + a_{23}mn) + \lambda, \qquad \text{(P.I.10)}$$

the quantities a_{11}, a_{12}, ... representing the sums of the squares and of the products of the coordinates x_i, y_i, and z_i.

The necessary conditions for an extremum are now simply

$$\partial \bar{F}/\partial l = 0, \qquad \partial \bar{F}/\partial m = 0, \qquad \partial \bar{F}/\partial n = 0, \qquad \text{(P.I.11)}$$

which gives

$$\begin{array}{llll} (a_{11} - \lambda) l + a_{12}m & + a_{13}n & = 0 \\ a_{12}l & + (a_{22} - \lambda) m + a_{23}n & = 0 \\ a_{13}l & + a_{23}m & + (a_{33} - \lambda) n = 0. \end{array} \qquad \text{(P.I.12)}$$

These equations represent a system of homogeneous linear equations. Apart from the trivial solution $l = m = n = 0$ they can be satisfied by a set of values of the unknowns only if the determinant of the coefficients is equal to zero,

$$\begin{vmatrix} a_{11} - \lambda & a_{12} & a_{13} \\ a_{12} & a_{12} - \lambda & a_{23} \\ a_{12} & a_{23} & (a_{33} - \lambda) \end{vmatrix} = 0. \qquad \text{(P.I.13)}$$

which is the cubic from which Lagrange's multiplyer λ can be computed.

The complete solution can quite easily be worked out using the mathematical developments shown in Exercise V, which is left to the reader.

Problem I.2. Formulae for Conversion of Equatorial into Galactic Coordinates

Deduce the formulae needed for conversion of equatorial coordinates into galactic coordinates.

Solution

The formulae for transformation of equatorial coordinates into galactic coordinates are most easily deduced applying the fundamental formulae of spherical trigonometry,

$$\sin a \cos B = \cos b \sin c - \sin b \cos c \cos A$$
$$\sin a \sin B = \sin b \sin A \qquad\qquad\qquad\qquad\qquad (\text{P.I.14})$$
$$\cos a = \cos b \cos c + \sin b \sin c \cos A$$

to the spherical triangle formed by the north celestial pole, the north galactic pole and the star. The following derivation, taken from the Introduction to Ohlsson's Conversion Tables, has the advantage of circumventing the explicit use of spherical trigonometry, now not so widely known as in former times.

Let in Figure I.3
EP represent the north pole of the equator 1900.0
GP, the north pole of the Galaxy having the position R.A. $= A$; Dec. $= D$ (1900.0)
Ω, the ascending node (and its right ascension) of the galactic plane on the equator for 1900.0, from which point the galactic longitudes are reckoned,
i, the inclination of the galactic plane to the equator for 1900.0.
Further we denote by
l, b the galactic longitude and latitude of a star, S, whose direction cosines in the galactic system of coordinates are ξ, η, ζ
α, δ, the right ascension and declination referred to the equinox 1900.0 of the star S, whose direction cosines in an equatorial system of coordinates, having the X-axis towards Ω, are x, y, z
φ the angle GP–S–EP or the galactic parallactic angle of the star S.

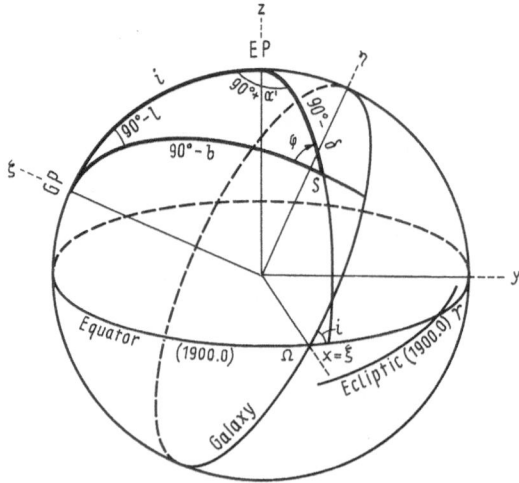

Fig. I.3.

Denoting for simplicity by α' the arc of the equator measured from Ω to the declination circle of the star, we have

$$\alpha' = \alpha - \Omega.$$

Introducing the values of the coordinates of the adopted position of the galactic pole

$$A = 12\,\text{h}\,40\,\text{m} = 190°; \qquad D = +28°$$

we obtain

$$\Omega = A + 90° = 280°; \qquad i = 90° - D = 62°$$

and

$$\alpha' = \alpha - 280° = \alpha + 80°.$$

According to the well-known formulae for the direction cosines we have

$$\begin{aligned}
\xi &= \cos b \cos l & x &= \cos \delta \cos \alpha' \\
\eta &= \cos b \sin l & y &= \cos \delta \sin \alpha' \\
\zeta &= \sin b & z &= \sin \delta
\end{aligned}$$
(P.I.15)

while, applying the relations between the direction cosines obtained by rotating the system of co-ordinates around the common axes of X and ξ through the angle i, we get, also introducing the numerical values of the coefficients,

$$\begin{aligned}
\xi &= x \\
\eta &= y \cos i + z \sin i = 0.46947\,y + 0.88295\,z \\
\zeta &= -y \sin i + z \cos i = -0.88295\,y + 0.46947\,z
\end{aligned}$$
(P.I.16)

and conversely

$$\begin{aligned}
x &= \xi \\
y &= \eta \cos i - \zeta \sin i = 0.46947\,\eta - 0.88295\,\zeta \\
z &= \eta \sin i + \zeta \cos i = 0.88295\,\eta + 0.46947\,\zeta
\end{aligned}$$
(P.I.17)

where (P.I.16) gives the transformation from equatorial coordinates to galactic coordinates.

If the galactic coordinates of several stars with independent positions are required, the above formulae are to be preferred. For the construction of the conversion tables another form of the formulae, however, will be found more convenient. Substituting the values of ξ, η, ζ, and x, y, z, given by (P.I.15) in the relations (P.I.16) and (P.I.17) we have

$$\begin{aligned}
\cos b \cos l &= \cos \delta \cos \alpha' \\
\cos b \sin l &= \cos i \cos \delta \cos \alpha' + \sin i \sin \delta \\
\sin b &= -\sin i \cos \delta \sin \alpha' + \cos i \sin \delta
\end{aligned}$$
(P.I.18)

and conversely

$$\begin{aligned}
\cos \delta \cos \alpha' &= \cos b \cos l \\
\cos \delta \sin \alpha' &= \cos i \cos b \sin l - \sin i \sin b \\
\sin \delta &= \sin i \cos b \sin l + \cos i \sin b
\end{aligned}$$
(P.I.19)

which can also be obtained by a direct application of the formulae of spherical trigonometry to the spherical triangle GP–S–EP.

To adapt the Equations (P.I.18) for the computation of the tabular values, we eliminate $\cos b$ from the first two equations by dividing the second by the first and introduce

$$\begin{aligned}
p &= \cos i = 0.46947 \\
q &= \sin i\,\text{tg}\,\delta = 0.88295\,\text{tg}\,\delta \\
r &= -\sin i \cos \delta = -0.88295 \cos \delta \\
s &= \cos i \sin \delta = 0.46947 \sin \delta.
\end{aligned}$$

Then the equations take the following form

$$\text{tg}\,l = p\,\text{tg}\,\alpha' + q \sec \alpha'$$
(P.I.20)
$$\sin b = r \sin \alpha' + s$$
(P.I.21)

which are the required expressions for transforming from equatorial coordinates into galactic co-ordinates. We observe that (P.I.21) fully determines b, also with respect to its sign. In the use of (P.I.20) we can fix the quadrant of l from the sign of $\text{tg}\,l$, observing that according to the first equation of (P.I.18), l and α' at the same time fall either in the first and fourth quadrant or in the second and third quadrant.

Problem I.3.   The Case when the Observer is not in the Galactic Plane: van Tulder's Method

J. J. van Tulder made, in 1942, a new determination of the galactic pole and of the distance of the Sun from the galactic plane. This determination was based on the investigation of the space distribution of several groups of Population I stars as well as other objects.

In his paper, published in the *Bulletin of the Astronomical Institutes of the Netherlands* (*Bull. Astron. Inst. Neth.* **9** (1942), 315) Van Tulder gives the following equation of condition on which the computations were based:

$$z = z_0 + r \sin b \cos \Delta + r \cos b \cos l \sin \Delta \cos l_0 + r \cos b \sin l \sin \Delta \sin l_0$$

in which

    r = distance of a star from the Sun,

    z = distance of a star from the true plane of symmetry,

    z_0 = distance of the Sun from the plane of symmetry, counted positive in the direction of the (true) north galactic pole.

    l, b are the galactic coordinates of the star in Ohlsson's system

    l_0 and $90° - \Delta$ are the galactic longitude and the galactic latitude, respectively, of the pole which is to be determined, referred to the old system of galactic coordinates.

Deduce Van Tulder's equation.

As to how this formula is to be used, we cite Van Tulder:

The solution is determined by the condition that Σz^2 must become a minimum. The equation in this form may be used only if accurate individual distances are known, which is only true for the group of δ Cephei variables nearer than 1000 pc. In all other cases the equation had to be written in the following form:

$$z/r = z_0/r + \sin b + (\cos b \cos l) X + (\cos b \sin l) Y, \qquad \begin{array}{l} X = \sin \Delta \cos l_0, \\ Y = \sin \Delta \sin l_0. \end{array}$$

The corresponding condition is now that $\Sigma (z/r)^2$ becomes a minimum. In order to prevent some interval obtaining too much weight, due to the often rather inhomogeneous distribution of stars, the stars have been combined in intervals of 30° longitude; the same weight has been given to each 30° interval irrespective of the number of stars contained in it. X, Y, and z_0/r were found from the twelve equations corresponding to the twelve intervals of longitude; z_0 was obtained from z_0/r by multiplying with r (the mean distance).

From X and Y it is easy to find Δ and l_0 which determine the new pole.

The remark concerning the distances does no more hold today. The concept of the equation of condition is explained in Appendix I.

Solution

Take a coordinate system having its origin in the Sun. Orient its axes as follows:

    The ξ-axis towards the point $l=0°$, $b=0°$

    the η-axis towards the point $l=90°$, $b=0°$

    the ζ-axis towards the North Galactic Pole,

corresponding to Ohlsson's system of galactic coordinates. By definition the new
z-axis passes through the Sun. Denote by λ, μ, ν, its direction cosines with respect to
the old axes. Then (Figure I.4)

$$\lambda = \sin \Delta \cos l_0 ; \qquad \mu = \sin \Delta \sin l_0 ; \qquad \nu = \cos \Delta .$$

Denote by z_0 the z-coordinate of the Sun in the new coordinate system. Applying the

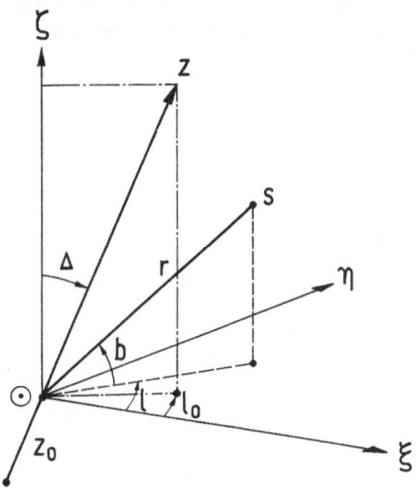

Fig. I.4.

well-known formula giving the coordinates in a rotated coordinate system one immediately gets

$$z - z_0 = \lambda \xi + \mu \eta + \nu \zeta .$$

Observing that

$$\xi = r \cos b \cos l, \qquad \eta = r \cos b \sin l, \qquad \zeta = r \sin b$$

one has

$$z = z_0 + r \cos b \cos l \sin \Delta \cos l_0 + r \cos b \sin l \sin \Delta \sin l_0 + r \sin b \cos \Delta$$

which is Van Tulder's formula.

References

[1] Bečvář, A.: 1960, *Atlas Coeli II, Katalog 1950*, Academy of Sciences of Czechoslovakia, Prague.
[2] Sawyer Hogg, H.: 1959, in S. Flügge (ed.), *Encyclopaedia of Physics*, Vol. LIII, Springer-Verlag,
    Berlin, p. 129.
[3] Alter, G., Ruprecht, J., and Vanýsek, V.: 1958, *Catalogue of Star Clusters and Associations*,
    Prague.
[4] Roberts, M. C.: 1962, *Astron. J.* **67**, 79.
[5] Smith, L. F.: 1968, *Monthly Notices Roy. Astron. Soc.* **138**, 109.

[6] Kukarkin, B. V., Parenago, P. P., Yefremov, Y. I., and Cholopov, P. N.: 1958, *General Catalogue of Variable Stars*, Academy of Sciences of the U.S.S.R., Moscow, 1958.
[7] Kraft, R. P., and Schmidt, M.: 1963, *Astrophys. J.* **137**, 249.
[8] *Trans. IAU*, **IV** (1932), p. 287.
[9] Ohlsson, J.: 1932, *Ann. Obs. Lund* **3**, Lund.
[10] Allen, C. W.: 1955, *Astrophysical Quantities*, 1st ed., The Athlone Press, London.
[11] McCuskey, S. W.: 1965, in A. Blaauw and M. Schmidt (eds.) *Galactic Structure*, Chicago University Press, Chicago, p. 1, esp. § 3.

Note Added in Proof. There is a new edition (1970) of reference [3], published by the Academy of Sciences, Budapest, and also a new (third) edition of reference [6], published by Nauka, Moscow, 1969–1971.

TABLE I.1[a]

A list of classical Cepheids

Star	α	δ	Star	α	δ
FM Cas	00^{h}09^{m}11^s	$+55°41'.8$	AD Gem	06^{h}37^{m}10^s	$+21°02'.2$
SY Cas	00 09 51	$+57$ 52 .1	TX Mon	06 45 50	-01 19 .0
DL Cas	00 24 24	$+59$ 40 .0	SZ Mon	06 46 24	-01 15 .2
AP Cas	00 27 35	$+62$ 21 .2	ζ Gem	06 58 11	$+20$ 43 .0
XY Cas	00 44 06	$+59$ 34	TV CMa	07 04 42	-13 37
VW Cas	00 59 35	$+61$ 13 .1	RW CMa	07 08 47	-18 33 .8
UZ Cas	01 06 22	$+60$ 40 .9	RY CMa	07 11 56	-11 18 .6
BP Cas	01 08 24	$+65$ 04 .7	SS CMa	07 21 58	-25 03 .6
RW Cas	01 30 43	$+57$ 14 .9	VW Pup	07 27 25	-19 56 .5
AY Cas	01 31 58	$+64$ 28 .8	VX Pup	07 28 19	-21 43 .0
BY Cas	01 40 19	$+60$ 55 .3	X Pup	07 28 26	-20 41 .7
VV Cas	01 44 17	$+59$ 23 .5	VZ Pup	07 34 34	-28 16 .2
VX Per	02 00 50	$+57$ 58 .0	WW Pup	07 37 35	-20 53 .9
UX Per	02 06 06	$+57$ 36 .6	WX Pup	07 37 50	-25 38 .5
VY Per	02 20 19	$+58$ 28 .0	AD Pup	07 43 53	-25 19 .7
DW Per	02 26 41	$+56$ 35 .3	BM Pup	07 45 59	-28 00 .0
UY Per	02 27 10	$+58$ 23 .5	WY Pup	07 53 50	-23 46 .4
DF Cas	02 37 00	$+61$ 04	AP Pup	07 54 17	-39 51 .2
SU Cas	02 43 03	$+68$ 28 .4	AQ Pup	07 54 19	-28 51 .6
AB Cam	03 38 00	$+58$ 28 .2	WZ Pup	07 56 07	-23 25 .7
RW Cam	03 46 11	$+58$ 21 .3	BN Pup	08 02 20	-29 48 .5
RX Cam	03 56 42	$+58$ 23 .0	AX Vel	08 07 45	-47 24 .1
TV Cam	03 59 02	$+60$ 02 .8	AT Pup	08 08 39	-36 38 .6
SX Per	04 10 13	$+41$ 29 .0	RS Pup	08 09 14	-34 16 .6
AS Per	04 12 22	$+48$ 42 .6	V Car	08 26 41	-59 47 .3
AW Per	04 41 05	$+36$ 32 .6	RZ Vel	08 33 34	-43 45 .9
SV Per	04 42 46	$+42$ 06 .8	T Vel	08 34 26	-47 00 .7
RX Aur	04 54 28	$+39$ 48 .7	SW Vel	08 40 21	-47 02 .5
SY Aur	05 05 31	$+42$ 42 .5	SX Vel	08 41 31	-45 58 .8
Y Aur	05 21 31	$+42$ 21 .2	ST Vel	08 41 50	-50 11 .8
β Dor	05 32 45	-62 33 .3	BG Vel	09 05 03	-51 01 .9
RZ Gem	05 56 35	$+22$ 14 .0	V Vel	09 19 15	-55 32 .0
AA Gem	06 00 22	$+26$ 20 .3	AE Vel	09 33 29	-52 35 .0
CS Ori	06 01 51	$+11$ 09 .4	l Car	09 42 30	-62 02 .8
RS Ori	06 16 31	$+14$ 43 .5	CN Car	10 12 02	-57 40 .5
T Mon	06 19 49	$+07$ 08 .4	RY Vel	10 16 56	-54 49 .1
RT Aur	06 22 08	$+30$ 33 .3	AQ Car	10 17 58	-60 34 .1
BB Gem	06 28 57	$+13$ 09 .5	UW Car	10 23 15	-59 09 .6
W Gem	06 29 14	$+15$ 24 .5	YZ Car	10 24 38	-58 50 .3
CV Mon	06 31 50	$+03$ 08 .9	UX Car	10 25 26	-57 06 .1

[a] The equatorial coordinates, taken from the *General Catalogue of Variable Stars* [6], are for 1900.0.

Table I.1 (continued)   A list of classical Cepheids

Star	α	δ	Star	α	δ
UY Car	$10^h28^m31^s$	$-61°16'.1$	R TrA	$15^h10^m49^s$	$-66°07'.8$
CR Car	10 29 09	$-58\ 00 .4$	U Nor	15 34 37	$-54\ 59 .3$
Y Car	10 29 25	$-57\ 59 .0$	SY Nor	15 46 57	$-54\ 16 .1$
XX Vel	10 32 14	$-55\ 31 .4$	S TrA	15 52 12	$-63\ 29 .5$
UZ Car	10 32 38	$-60\ 29 .6$	RS Nor	15 57 24	$-53\ 38 .5$
CT Car	10 33 17	$-61\ 03 .7$	GU Nor	16 07 09	$-53\ 05 .0$
EY Car	10 38 38	$-60\ 38 .6$	S Nor	16 10 34	$-57\ 39 .2$
VY Car	10 40 35	$-57\ 02 .4$	RV Sco	16 51 47	$-33\ 27 .2$
SV Vel	10 40 55	$-55\ 45 .9$	BF Oph	16 59 53	$-26\ 26 .6$
SX Car	10 42 07	$-57\ 01 .2$	V 482 Sco	17 24 13	$-33\ 31 .9$
WW Car	10 47 36	$-58\ 51 .3$	X Sgr	17 41 16	$-27\ 47 .6$
DY Car	10 48 38	$-59\ 59 .6$	V500 Sco	17 42 12	$-30\ 26 .4$
FZ Car	10 49 59	$-58\ 40 .0$	RY Sco	17 44 16	$-33\ 40 .5$
WZ Car	10 51 20	$-60\ 24 .4$	W Sgr	17 58 38	$-29\ 35 .1$
XX Car	10 53 21	$-64\ 36 .0$	AP Sgr	18 06 58	$-23\ 08 .5$
U Car	10 53 44	$-59\ 11 .8$	WZ Sgr	18 11 06	$-19\ 06 .6$
CY Car	10 53 50	$-60\ 12 .6$	Y Sgr	18 15 30	$-18\ 54 .3$
FN Car	10 57 08	$-59\ 34 .6$	AY Sgr	18 17 26	$-18\ 37 .4$
FO Car	10 57 29	$-61\ 45 .1$	XX Sgr	18 18 57	$-16\ 51 .0$
XY Car	10 58 19	$-63\ 43 .5$	X Sct	18 25 41	$-13\ 10 .6$
HK Car	10 59 38	$-60\ 06 .3$	U Sgr	18 26 00	$-19\ 11 .7$
XZ Car	11 00 06	$-60\ 26 .4$	Y Sct	18 32 36	$-08\ 27 .3$
ER Car	11 05 24	$-58\ 17 .7$	RU Sct	18 36 39	$-04\ 12 .4$
IT Car	11 07 56	$-61\ 12 .7$	TY Sct	18 36 50	$-04\ 23 .3$
FR Car	11 10 01	$-59\ 30 .5$	Z Sct	18 37 36	$-05\ 55 .1$
AY Cen	11 20 37	$-60\ 11 .1$	SS Sct	18 38 18	$-07\ 49 .8$
AZ Cen	11 20 45	$-60\ 49 .2$	V350 Sgr	18 39 20	$-20\ 45 .0$
UZ Cen	11 36 15	$-62\ 08 .3$	YZ Sgr	18 43 42	$-16\ 50 .1$
RT Mus	11 39 50	$-66\ 45 .0$	BB Sgr	18 45 04	$-20\ 24 .7$
BK Cen	11 44 24	$-62\ 31 .2$	FF Aql	18 53 48	$+17\ 13 .6$
TZ Mus	11 46 04	$-64\ 35 .0$	V336 Aql	18 56 13	$+00\ 00 .3$
UU Mus	11 47 23	$-64\ 50 .8$	SZ Aql	18 59 35	$+01\ 09 .4$
S Mus	12 07 24	$-69\ 35 .7$	V496 Aql	19 02 57	$-07\ 35 .5$
AD Cru	12 07 42	$-61\ 32 .4$	TT Aql	19 03 09	$+01\ 08 .5$
SU Cru	12 12 50	$-62\ 43 .5$	FM Aql	19 04 32	$+10\ 23 .5$
T Cru	12 15 53	$-61\ 43 .6$	FN Aql	19 07 48	$+03\ 23 .4$
R Cru	12 18 08	$-61\ 04 .5$	U Aql	19 23 58	$-07\ 15 .0$
VW Cru	12 27 36	$-62\ 57 .2$	SU Cyg	19 40 48	$+29\ 01 .4$
AG Cru	12 35 41	$-59\ 14 .7$	η Aql	19 47 23	$+00\ 44 .9$
R Mus	12 35 58	$-68\ 51 .5$	SV Vul	19 47 25	$+27\ 12 .3$
X Cru	12 40 33	$-58\ 34 .7$	S Sge	19 51 29	$+16\ 22 .2$
S Cru	12 48 27	$-57\ 53 .3$	GH Cyg	19 55 06	$+29\ 10 .9$
KN Cen	13 29 35	$-64\ 02 .8$	CD Cyg	20 00 37	$+33\ 49 .7$
XX Cen	13 33 46	$-57\ 06 .4$	V402 Cyg	20 05 25	$+36\ 51 .5$
V Cen	14 25 23	$-56\ 26 .7$	MW Cyg	20 08 27	$+32\ 34 .5$

Table 1.1 (continued)   A list of classical Cepheids

Star	α	δ	Star	α	δ
SZ Cyg	20ʰ29ᵐ38ˢ	+46°15′.6	CR Cep	22 42 30	+ 58 54 .9
X Cyg	20 39 29	+ 35 13 .6	V Lac	22 44 33	+ 55 47 .6
BZ Cyg	20 42 34	+ 44 56 .5	X Lac	22 44 58	+ 55 54 .0
VX Cyg	20 53 34	+ 39 47 .5	SW Cas	23 02 54	+ 58 00 .8
TX Cyg	20 56 26	+ 42 12 .4	CH Cas	23 17 36	+ 62 12 .5
VY Cyg	21 00 27	+ 39 34 .5	CY Cas	23 24 44	+ 62 49 .3
DT Cyg	21 02 18	+ 30 47 .0	RS Cas	23 32 36	+ 61 52 .5
V459 Cyg	21 07 31	+ 48 44 .0	DW Cas	23 34 00	+ 58 48
V386 Cyg	21 10 52	+ 41 18 .1	CZ Cas	23 34 46	+ 61 49 .6
V532 Cyg	21 16 53	+ 45 02 .6	CD Cas	23 40 17	+ 62 27 .0
MZ Cyg	21 17 53	+ 37 02 .3	RY Cas	23 47 10	+ 58 11 .1
VZ Cyg	21 47 41	+ 42 39 .9	DD Cas	23 52 19	+ 62 09
CP Cep	21 54 27	+ 55 41 .2	CEa Cas	23 53 07	+ 60 39 .2
BG Lac	21 56 21	+ 42 57 .9	CEb Cas	23 53 07	+ 60 39 .2
Y Lac	22 05 13	+ 50 33 .3	CF Cas	23 53 15	+ 60 39 .7
DF Lac	22 17 28	+ 54 01 .7	CG Cas	23 55 54	+ 60 24 .7
AK Cep	22 25 11	+ 57 42 .0	IX Cas	23 59 42	+ 49 40 .7
δ Cep	22 25 27	+ 57 54 .2			
Z Lac	22 36 55	+ 56 18 .4			
RR Lac	22ʰ37ᵐ28ˢ	+ 55°54′.6			

EXERCISE II

THE APPARENT DISTRIBUTION AND THE COLORS OF
THE GLOBULAR CLUSTERS

1. The purpose of the present exercise is twofold. First, we intend to show how from a quite simple analysis of the apparent distribution of the globular clusters one can infer the existence of a center in our stellar system as well as of the absorption of star light in the vicinity of the galactic equator. Moreover we shall be able to derive approximate values for the coordinates of this center. Second, we shall prove that there is a statistical relation between the color of a globular cluster and its distance from the galactic equator. This effect can be interpreted by differential interstellar absorption in a thin layer of absorbing material distributed in the vicinity of the galactic plane. By using a simple model it is possible to get an estimate of the thickness of this layer.

Some of the questions which will be dealt with in this exercise in a simplified and condensed form, are related to the problem of the redefinition of the system of galactic coordinates. For the reader's convenience this problem will briefly be discussed in Section 3, which also contains the definition of the new system of galactic coordinates adopted at present.

Why shall we choose the globular clusters? One reason is that in many respects, and in particular by their very weak concentration towards the galactic equator, the galactic clusters (typical Population II) represent just the opposite to the highly flattened classes of galactic objects considered in the first exercise. But even when compared to other Population II objects, the globular clusters deserve special attention and this for the following reasons. First, it seems that our knowledge of the system of globular clusters is as complete as materially possible. In fact, the number of known objects of this kind has hardly increased during the last decades. We can therefore safely assume that the volume of space around the Sun, defined by the penetrating power of our instruments, has been exhaustively searched for globular clusters. Certainly we do not already know all the globular clusters existing in our Galaxy, so that those which are listed in our catalogues represent only a sample taken from a wider population. However, as far as we can guess, this sample contains a sufficiently large proportion of all globular clusters. Moreover, and this is a particularly important point, in the search for globular clusters no particular region of the sky nor any particular subgroup has been favored. In this sense it can be stated that the data at our disposal are not distorted or biased by selection effects. In other words, notwithstanding the fact that our knowledge of the system of the globular clusters is limited to a sample, this sample can be considered as sufficiently representative for the whole.

It is important to realise that such a favorable situation does not always occur. Take, for example, the planetary nebulae. By comparing the catalogues of the planetary nebulae published between 1948 and 1967 one sees that during this time the number of known objects of this kind has risen from somewhat less than 300 to more than 1000. Moreover some regions of the sky have been observed with more powerful instruments, and in these more objects have been recorded. In other words the corresponding data are not statistically homogeneous, but instead distorted by selection effects. Before using such data for statistical investigation, they must be corrected for selection effects.

2. Let us begin with an investigation of the apparent distribution of the globular clusters. First we shall make use of a catalogue of these objects compiled by Mrs. H. Sawyer Hogg [1]. It contains 118 entries, but according to H. C. Arp the objects NGC 2158 and Russ should be omitted.

The overwhelming majority of clusters listed in this catalogue can be identified by their NGC or IC numbers. Now if one remembers that Dreyer's *New General Catalogue* (abbreviation: NGC) was published in 1888, and its two supplements (*The Index Catalogues* I and II; abbreviation: IC) in 1895 and 1908 respectively, and if, moreover, the enormous progress in observing techniques since that time is taken into account, this fact illustrates quite well our preceding statement about the completeness of our knowledge of the globular clusters in our part of the Galaxy.

In Mrs. Sawyer Hogg's catalogue the globular clusters are arranged in order of their right ascensions for 1950.0. Even at first glance one can see that the distribution of the globular clusters in right ascension is far from uniform. However this distribution follows a quite regular course. In fact the number of entries per interval of one hour in right ascension rises to a broad but very pronounced maximum, situated somewhere between 17 h and 19 h, and then falls off. The zone between 17 h and 19 h contains more than 40% of all the globular clusters listed, the hemisphere between 15 h and 21 h more than 70% of them.

Although quite conspicuous in the system of equatorial coordinates, this peculiar feature in the apparent distribution of globular clusters, as well as other ones, will be revealed much more clearly when galactic coordinates are used. They are given in the fifth and the sixth column of the catalogue. The galactic longitude is denoted by λ, the latitude by β. Both refer to Ohlsson's system of coordinates.

First examine the distribution of the globular clusters in galactic longitude. For that purpose count the number of clusters per interval of 10° in galactic longitude, from the galactic longitude 0° on. Note that a convention is needed for those clusters which have longitude equal to a multiple of 10, e.g. equal to 170°. They can be divided half and half between the two adjacent intervals. Give the result in the form of a histogram (Figure II.1) with the galactic longitude on the horizontal axis and the number $n(l)$ of clusters per interval of 10° on the vertical axis. Trace through the midpoints indicated on our schematical Figure II.1 by small circles a continuous curve and estimate the longitude of its maximum l_0. Note that in the vicinity of this point the curve is quite symmetric.

Investigate now the distribution, in galactic latitude, of those clusters which have longitudes within the interval $l_0+90°$ and $l_0-90°$. As before count the number of

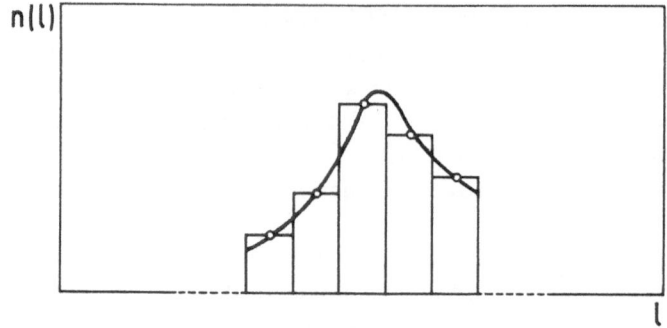

Fig. II.1.

clusters per interval of 10° in galactic latitude. For reasons which will appear later take an interval centered on the galactic equator $(-5 < b < +5°)$, the other one being placed symmetrically and bounded by galactic parallels ten degrees apart, up to $\pm 85°$ (obviously the cluster NGC 288 will not be taken into account).

As before, give the distribution in galactic latitude in the form of a histogram. Note however that in the present case the counts in different intervals are not directly comparable, as the zones to which they refer have different areas. Therefore the counts must be reduced to equal areas, e.g. to unit area on the celestial sphere. Now the area of the zone bounded by the parallels b_1 and b_2 is equal to $2\pi (\sin b_2 - \sin b_1)$, the radius of the sphere being taken as unity. Remember, furthermore, that the counts refer only to a half of each zone, as only clusters within the interval $l_0 - 90°$ to $l_0 + 90°$ have been taken into account. Finally, let n' be the number of clusters counted in the interval between b_1 and b_2. Then the number of clusters per unit area, n, will be given by

$$n = n'/\pi \cdot (\sin b_2 - \sin b_1). \tag{II.1}$$

Make the histogram n versus the galactic latitude b.

The existence of the galactic concentration will clearly be indicated by the rise of n when approaching the galactic equator. However on the equator itself a conspicuous depression will appear. The number of clusters effectively present in the equatorial band, i.e. between the latitudes $-5°$ and $+5°$, falls far below that expected from the steep rise on both sides of the equator.

It is proposed that the more interested reader should make in the same way another n versus b histogram adopting an interval only 4° wide in galactic latitude. As the effect we are interested in occurs in the equatorial band, he can count only those globular clusters falling within the parallels, say, $-22°$ to $-18°$, $-18°$ to $-14°$, ..., $-2°$ to $+2°$, ..., $+18°$ to $+22°$. As it will be seen the equatorial depression will be such more pronounced on the histogram made with a narrower interval. The choice of a wider interval obviously has a smoothing effect, which means that some important details may be lost. On the other hand if the interval is too narrow, the statistical fluctuations will become too prominent. In other words the choice of a suitable interval width may not always be as simple as it seems.

In order to get a better insight into the phenomenon disclosed by the n versus b diagram, make a chart of all the globular clusters using the same projection as in the preceding exercise, but now with galactic coordinates. See Figure II.2. Do not trace the straight line which corresponds to the galactic equator. As you will see, the globular clusters are practically absent in a narrow band along the galactic equator. However

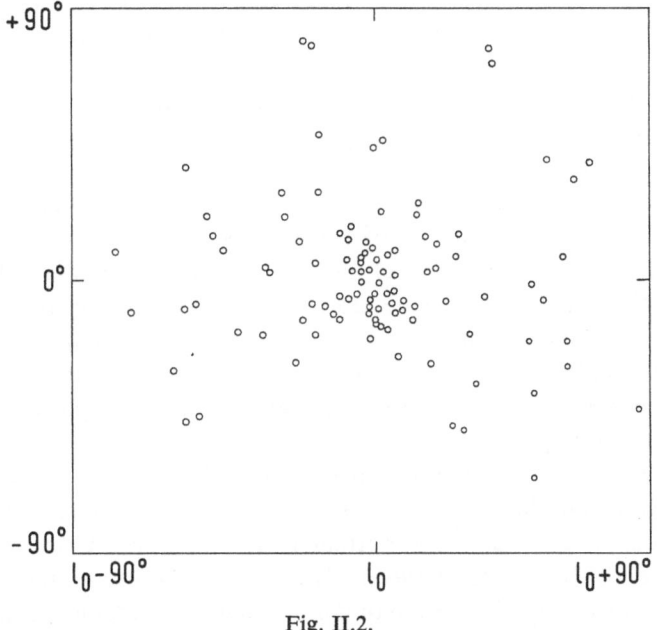

Fig. II.2.

this cannot be considered as a feature proper to the system of globular clusters as it is found not only in the apparent distribution of many other galactic objects, but also in the distribution of external galaxies (the zone of avoidance). This effect is due to the absorption of light in a thin layer of interstellar matter concentrated in the vicinity of the galactic plane.

Due to the peculiar distribution of the globular clusters in galactic latitude it is impossible to locate on the histograms n versus b the maximum, which is obviously hidden behind the absorbing layer. However we certainly can assume that its latitude corresponds to the mean value of the latitudes of the globular cluster concerned. Compute this mean value b_0

Summarising, we can say that the analysis of the apparent distribution of the globular clusters has revealed the existence of a point of accumulation. This fact can be interpreted by assuming that, in space, the globular clusters are distributed more or less symmetrically around a center lying far from the Sun and very approximately in the galactic plane (compare your value for b_0). Now this same point reappears in the distribution of many more groups of objects exhibiting a moderate or weak galactic concentration. It also plays an important role in the kinematics of our stellar system (see Exercises IV and VIII). Convergent evidence indicates that it shall be identified with the center of the Galaxy. Compare your values l_0, b_0 to the coordinates of the Galactic center which in Ohlsson's system are

$$l = 328°, \quad b = -1°.$$

The existence of a well defined center represents another essential characteristic of our Galaxy. Let us consider only the bearing which this fact has on the definition of the plane of symmetry of our stellar system.

The definition of this plane as previously adopted, was based only on the distribution of objects strongly concentrated towards the Milky Way. Now in the distribution of these objects along the galactic equator the point of accumulation so conspicuous in the distribution of weakly concentrated objects does not even appear. This means that when defining the plane of symmetry of our stellar system we in fact have ignored the direction towards its center. In view of our present results and interpretations we obviously shall stipulate that this plane has to pass through the center. Therefore the redefinition of the plane of symmetry seems well indicated. Moreover it is evident that now we can also make a better choice for the zero point of galactic longitudes, as it is obvious that the direction towards the galactic center represents a physically meaningful preferential direction.

The very interesting problem of the redefinition of the system of galactic coordinates cannot be dealt with in the present exercise. For an exhaustive discussion we refer the interested reader to the series of important papers on this subject published in the *Monthly Notices of the Royal Astronomical Society* [2]. This series represents the final report of the former Commission 33b of the International Astronomical Union which at the IX General Assembly of the IAU (Dublin, 1955) was appointed "to investigate the desirability of the revision of the position of the galactic pole and of the zero of galactic longitude". However in view of the importance of this question we shall give, in the following section, some excerpts from the definition and the announcement of the new system of galactic coordinates adopted at present and add some brief comments.

3. The X General Assembly of the IAU (Moscow, 1958) has endorsed the following resolution passed at a previous Joint Meeting of the IAU Commissions 33 (The Structure and Dynamics of the Stellar System) and 40 (Radio Astronomy):

"(a) That a standard system of galactic coordinates be adopted for which the pole is based primarily on the distribution of neutral hydrogen in the inner parts of the galactic system.

(b) That a zero of longitude be chosen near the longitude of the galactic nucleus, that the longitude be counted from 0° to 360°, in the same direction as in the current system, and that the latitude be counted in the conventional manner from $-90°$ through 0° to $+90°$.

(c) That Commission 33b be authorized to define the exact values of the coordinates of the pole and the longitude immediately after the final reduction of the relevant observation is finished." [2, p. 127.]

In its final report referred to above, this Commission announced the new system in the following way [2, p. 130]:

"(a) The new galactic north pole lies in the direction

$$\alpha = 12\,\text{h}\,49\,\text{m} \qquad \delta = +27.4° \text{ (equinox 1950.0)}$$

(b) The new zero of galactic longitude is the great semi-circle originating at the new north galactic pole at the position angle

$$\theta = 123°$$

with respect to the equatorial pole for 1950.0" ...

As to the notations to be used, "at the Moscow General Assembly it was decided that the symbols l, b, should be retained for galactic longitude and galactic latitude respectively. Commission 33b suggests that, during the transition period, the symbols l^I, b^I, should be used for the old system and l^{II}, b^{II}, for the new one. Apart from these subscripts, it should be made quite clear whether the galactic coordinates used in any publication are based on the old or the new system."

This transition period is now over. At the XIV General Assembly of the International Astronomical Union (Brighton, 1970) it was decided "that the superscript II be omitted from the new galactic coordinates: henceforth it will simply be l, b; the old coordinates retain the superscript I. It is desirable that authors indicate in their papers that l, b refer to the new galactic coordinates".

From now on we shall make use of the new galactic coordinates and adopt for them the notation l, b.

Exhaustive conversion tables have been computed at, and published by, the Lund Observatory, Sweden [3].

Let us briefly comment on this resolution. Although for some years it has been known that the position of the north galactic pole as adopted in the old, i.e. Ohlsson's system was probably in error by one degree at least, "the need for a new pole was probably not very strongly felt at this time owing to the fact that the scatter of poles derived from various groups of objects was of a similar order to, or larger than, the deviation". Also by optical investigations the direction towards the galactic center could only be determined with insufficient accuracy. It was not until the advent and prodigious development of radio-astronomical methods, that a redefinition of the galactic coordinate system became imperative and feasible.

As far as the definition of the fundamental plane is concerned, "observations at 21 cm have shown that the neutral hydrogen layer is exceedingly flat over a region of the Galaxy within 7 kpc from the center", so flat indeed as to be indistinguishable from a plane ... "The high degree of flatness of the inner region must be related to important dynamical properties of the whole Galaxy giving special significance to the mean plane of the matter in that region", which plane "can be regarded as the principal plane of the Galaxy" [2, p. 132]. Beyond 7 kpc the neutral hydrogen layer is distorted and deviates from the plane defined above. Other radio-astronomical as well as optical investigations give strong support to the choice of the new galactic pole. However as they are of lower accuracy, the position of the principal plane, or what amounts to the same, the position of the new galactic pole, is primarily based on neutral hydrogen distribution.

Let us remark that in all the papers which form the final report from which these quotations have been taken, the probably too low value of 8.2 kpc for the Sun's distance from the galactic center was adopted.

Turning now to the question concerning the choice of the zero of longitude it is important that in the presumed direction of the galactic center a peculiar radio source, Sagittarius A, has been found. This source can almost certainly be identified with the galactic nucleus, for "it would be an extremely improbable coincidence if this unique source should accidentally lie within 0.1° of the center without being connected with it" (Oort, Rougoor). Its position can be determined by radio-astronomical methods with a precision far exceeding that with which the galactic center can be located by any other means. Within observational errors this source lies well in the principal plane of the Galaxy. The positional angle, with respect to the equatorial pole for 1950.0, of the great semicircle originating from the new north galactic pole and drawn through the assumed position of that source is equal to 123°00′30″, a value which may conveniently be rounded to 123° as proposed in the announcement of the new galactic coordinate system.

Note that it is by no means selfevident that the geometrical and kinematical center of a system of discrete bodies shall be materialized by a physical object – the galactic nucleus. The fact that such an object exists and that, moreover, it has peculiar properties must be considered as especially significant.

Let us finally remark that the position of the principal plane has been determined by making use of Van Tulder's method mentioned in Exercise I, Problem 2. For the height of the Sun above the plane a value of 4 pc ± 12 pc (estimated probable error) has been found. It can therefore be said that, according to the best results available at present, the Sun practically lies in the principal plane of the Galaxy.

4. Let us now resume our discussion and consider the second point mentioned in the introduction.

The peculiar form of the n versus b histogram clearly indicates the presence of absorbing material in the vicinity of the galactic plane. We shall now show the absorption it produces is selective, i.e. that it varies with the wavelength, or with the region of the spectrum concerned, affecting more strongly radiation of shorter wavelength.

First let us adopt, for the layer of absorbing material, the simplest possible model and assume that it is thin, plane-parallel, homogeneous, and symmetric about the galactic plane (Figure II.3). Let g be a globular cluster and S the Sun. Then b is the galactic latitude of the cluster. Under the assumptions made, the quantity of absorbing material

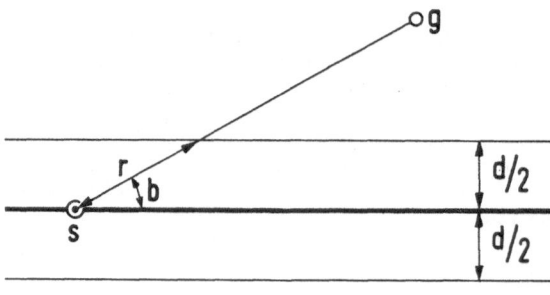

Fig. II.3.

traversed by radiation coming from g depends only on the galactic latitude b, or more precisely, on its absolute value $|b|$. Now it is known from elementary physics that, after traversing a slab of homogeneous absorbing material of thickness r, the incident intensity I_0 of a (monochromatic) ray is reduced and becomes

$$I = I_0 \, e^{-\alpha r} \tag{II.2}$$

where α is the absorption coefficient of the material. Expressing the relative decrease in intensity in magnitudes we have

$$\Delta m_{\text{abs}} = -2.5 \log(I/I_0) = 1.086\alpha r = ar. \tag{II.3}$$

In other words the absorption by a homogeneous layer produces an increase in the observed magnitude which is proportional to the thickness of the material interposed between the object and the observer. For an object imbedded in the absorption layer, e.g. for a star of a galactic star cluster at low galactic latitude the thickness will simply be equal to the distance of the object. For globular clusters which are assumed to be at large distances and outside the layer, it will be equal to

$$(d/2): \sin|b| = (d/2)\csc|b| \tag{II.4}$$

where d is the thickness of the layer in the direction perpendicular to the galactic plane (see Figure II.3). Therefore for a globular cluster seen at galactic latitude b, the increase in magnitude due to absorption will be

$$\Delta m_{\rm abs} = \tfrac{1}{2}ad \csc|b|. \tag{II.5}$$

The absorption coefficient a is usually expressed in magnitudes per kiloparsec.

Suppose now that we have measured the magnitudes of a globular cluster which correspond to two different regions of the spectrum, e.g. its B and V magnitudes. Let B_0 and V_0 be the values as they would be found if there were no absorption. Denote by a_B and a_V respectively the values of the absorption coefficient for the regions of the spectrum to which these magnitudes pertain. Notice that, due to the selective character of the absorption we must always specify the region of the spectrum to which it refers. Then the observed magnitudes of a globular cluster at a latitude b will be given by

$$\begin{aligned}
B_{\rm obs} &= B_0 + \tfrac{1}{2}a_B \cdot d \cdot \csc|b| \\
V_{\rm obs} &= V_0 + \tfrac{1}{2}a_V \cdot d \cdot \csc|b|.
\end{aligned} \tag{II.6}$$

The difference $(B-V)_{\rm obs}$ of the observed magnitudes is the observed (or, as it is sometimes called, the apparent) color index; the difference $(B-V)_0$ defines the intrinsic color index of the cluster. From (II.6) we get

$$(B - V)_{\rm obs} = (B - V)_0 + \tfrac{1}{2}\Delta a_{B,V}d \csc|b| \tag{II.7}$$

where $\Delta a_{B,V} = a_B - a_V$ is the difference of the absorption coefficients for the two spectral regions concerned, or, as it is usually called, the coefficient of differential absorption for these two regions of the spectrum. If the value of the absorption coefficient a_B which corresponds to the B magnitudes i.e. to the blue region of the spectrum is larger that the value a_V corresponding to the yellow region, the effect of the differential absorption will be to increase the color index, and the radiation will become redder. The difference $(B-V)_{\rm obs} - (B-V)_0$ is the color excess, E_{B-V}.

Let us now adopt the somewhat rough but useful assumption that statistically, or in the mean, all the globular clusters have the same intrinsic color index. Then by Equation (II.7) we shall expect that, in the mean, the observed color indexes $(B-V)_{\rm obs}$ of the globular clusters shall linearly depend on csc $|b|$.

Notice that instead of the standard color indexes $B-V$ we can also use other ones, referred to other regions of the spectrum. In Mrs. H. Sawyer Hogg's list the color indexes C_2 are those used by Stebbins and Whitford in their pioneering investigations of the colors of globular clusters [4]. They correspond to the differences of magnitudes measured in the regions of the spectrum centered on the wavelengths 4340 Å and 4670 Å respectively.

In order to verify the existence of this reddening we need a list of globular clusters with known galactic latitudes and measured color indexes.

The new galactic latitudes can be taken from a list of globular clusters compiled by H. C. Arp [5] which also contains many more useful data on these objects. The values of the (apparent) color indexes given by Arp have been derived from measurements made by G. E. Kron and N. U. Mayall [6] in the P, V photometric system. Another, even more complete, source of data on color indexes is S. van den Bergh's paper on the

UBV Photometry of Globular Clusters [7]. Table II.1 at the end of this Exercise
contains the color indexes given by Van den Bergh.

Make a list of globular clusters with known color indexes. The list shall contain the
NGC (or other) number of the cluster, its galactic latitude, the cosecant of the absolute
value of the galactic latitude (to one decimal place), and the measured, or apparent,
color index, as indicated below.

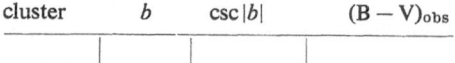

cluster	b	$\csc |b|$	$(B-V)_{obs}$

The values of the galactic latitude b given in Arp's list are rounded to the nearest
full degree. At low latitudes, i.e. for large values of $\csc |b|$ this will appreciably increase
the scatter of the points on your diagram. If desired more accurate values of the galactic latitude can be found from the equatorial coordinates given by Arp using conversion tables [3]. But even if this is not done, the result will be of the correct order of
magnitude.

Now make a diagram with the values of $\csc |b|$ on the horizontal axis and the
observed color index $(B-V)_{obs}$ on the vertical axis, Figure II.4. In spite of the very

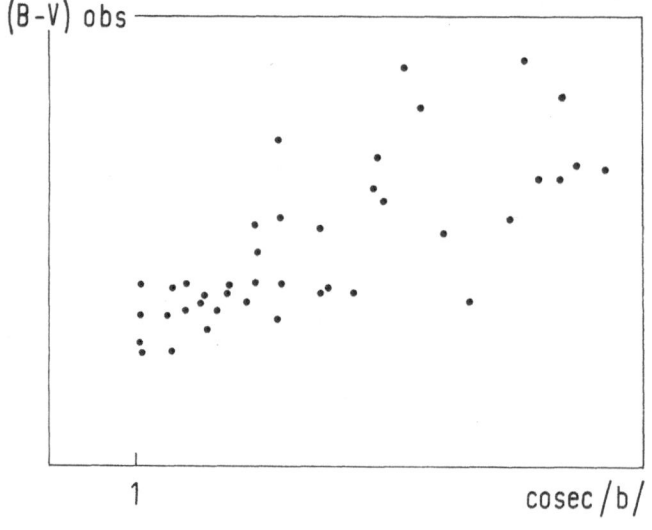

Fig. II.4.

large scatter of the points you will recognise the effect of the reddening, the points on
the right part of the diagram being systematically higher than those on the left part.

The large scatter of the points on your diagram reflects the imperfection of the basic
assumptions adopted. For it is certain that the globular clusters differ in color, and it
is well known that the absorbing (dust) layer has a patchy structure, characteristic of
all extreme Population I objects.

In order to get a clearer picture, instead of the individual values of the color indexes and cosecants, use averages taken in not too wide, suitably chosen intervals of csc $|b|$. From the diagram already made you can easily select those values of csc b for which the color indexes can be averaged.

Make a second diagram using the averaged values. Trace a straight line which best fits all the points and deduce the values of $(B-V)_0$ as well as of $\frac{1}{2}\Delta a_{B,V}d$. The intrinsic color index $(B-V)_0$ is determined by the point where the straight line intersects the vertical axis (which must be drawn through the zero on the horizontal axis). Obviously $\frac{1}{2}\Delta a_{B,V}d$ is the angular coefficient of the straight line.

According to C. W. Allen [8] the mean intrinsic color index is equal to $+0.6$, however compare Figure 4 in Arp's article [5]. The coefficient of csc $|b|$ usually adopted is about 0.06, see [5] and [7].

Notice that $\frac{1}{2}\Delta a_{B,V}d$ is the differential absorption, for the B and V regions, produced by a slab of absorbing material with a thickness equal to the half of the thickness of the absorbing layer (to prove this put $|b|=90°$ in Equation (II.7)). Therefore if the coefficient of differential absorption is known, one can determine the thickness of the absorbing layer. Now according to Allen [8], $a_V=3(a_B-a_V)$ and $a_V=2$ mag/kpc, so that we get $\Delta a_{B,V}=\frac{2}{3}$ mag/kpc. Make use of this value to derive an estimate of the thickness of the absorbing layer according to your determination. The effective thickness is of the order of 200 pc [9].

Until now we have considered the globular clusters as members of a homogeneous group. But is this in fact so? If he needs an answer, the reader may consult Arp's article [5]. But even with the data at his disposal he can try to find an answer. For this purpose mark on the chart of globular clusters (corresponding to our schematical Figure II.2) all the clusters of spectral type G. How are they distributed?

References

[1] Sawyer Hogg, H.: 1959, in S. Flügge (ed.), *Encyclopaedia of Physics*, Vol. LIII, Springer-Verlag, Berlin, p. 129.
[2] Blaauw, A., Gum, C. S., Pawsey, J. L., and Westerhout, G.: 1960, *Monthly Notices Roy. Astron. Soc.*, **121** 123.
[3] Torgard, I. (ed.): 1961, *Ann. Obs. Lund* **15–17** and Suppl. to **15–16**, Lund.
[4] Stebbins, J. and Whitford, A. E.: 1936, *Astrophys. J.* **84** 132.
[5] Arp, H. C.: in A. Blaauw and M. Schmidt (eds.), *Galactic Structure*, University of Chicago Press, Chicago, 1965, p. 401.
[6] Kron, G. E. and Mayall, N. U.: 1960, *Astron. J.* **65** 581.
[7] Van den Bergh, S.: 1967, *Astron. J.* **72**, 70.
[8] Allen, C. W.: 1963, *Astrophysical Quantities*, The Athlone Press, London.
[9] Voigt, H. H. (ed.): 1965, *Landolt-Börnstein, Numerical Data, Group VI*, Vol. 1. Springer-Verlag, Berlin, p. 659.

TABLE II.1

NGC	α	δ	l	b	B — V
104	$0^h21^m.9$	$-72°21'$	306°	$-45°$	0.86
288	0 50 .2	$-26\ 52$	(147)	-89	0.67
362	1 00 .6	$-71\ 07$	302	-47	0.75
1904	5 22 .2	$-24\ 34$	228	-29	0.60
2419	7 34 .8	$+39\ 00$	181	$+26$	0.69
2808	9 10 .9	$-64\ 39$	283	-11	0.93
3201	10 15 .5	$-46\ 09$	277	$+\ 9$	0.98
4147	12 07 .6	$+18\ 49$	251	$+77$	0.59
4590	12 36 .8	$-26\ 29$	299	$+37$	0.67
4833	12 56 .0	$-70\ 36$	304	$-\ 8$	0.96
5024	13 10 .5	$+18\ 26$	333	$+80$	0.63
5053	13 13 .9	$+17\ 57$	335	$+79$	0.64
5139	13 23 .8	$-47\ 03$	309	$+15$	0.79
5272	13 39 .9	$+28\ 38$	42	$+78$	0.69
5286	13 43 .0	$-51\ 07$	312	$+11$	0.89
5466	14 03 .2	$+28\ 46$	42	$+73$	0.76
5634	14 27 .0	$-05\ 45$	342	$+49$	0.67
5694	14 36 .7	$-26\ 19$	331	$+30$	0.72
5824	15 00 .9	$-32\ 53$	332	$+22$	0.74
5897	15 14 .5	$-20\ 50$	343	$+30$	0.73
5904	15 16 .0	$+02\ 16$	4	$+47$	0.72
5927	15 24 .5	$-50\ 29$	326	$+\ 5$	1.34
5946	15 31 .8	$-50\ 30$	327	$+\ 4$	1.20
5986	15 42 .8	$-37\ 37$	337	$+14$	0.89
6093	16 14 .1	$-22\ 52$	353	$+19$	0.86
6121	16 20 .6	$-26\ 24$	351	$+16$	1.04
6139	16 24 .3	$-38\ 44$	342	$+\ 7$	1.32:
6144	16 24 .2	$-25\ 56$	352	$+15$	0.95
6171	16 29 .7	$-12\ 57$	3	$+23$	1.11
6205	16 39 .9	$+36\ 33$	59	$+41$	0.67
6218	16 44 .6	$-01\ 52$	15	$+26$	0.85
6229	16 45 .6	$+47\ 37$	73	$+40$	0.76
6235	16 50 .4	$-22\ 06$	359	$+13$	0.90:
6254	16 54 .5	$-04\ 02$	15	$+23$	0.91
6266	16 58 .1	$-30\ 03$	353	$+\ 7$	1.14
6273	16 59 .5	$-26\ 11$	357	$+\ 9$	1.01
6284	17 01 .5	$-24\ 41$	358	$+10$	0.99
6287	17 02 .1	$-22\ 38$	0	$+11$	1.28
6293	17 07 .1	$-26\ 30$	357	$+\ 8$	0.99
6304	17 11 .4	$-29\ 24$	356	$+\ 5$	1.34
6316	17 13 .4	$-28\ 05$	357	$+\ 5$	1.31
6325	17 15 .0	$-23\ 42$	1	$+\ 8$	1.70
6333	17 16 .2	$-18\ 28$	5	$+10$	0.96
6341	17 15 .6	$+43\ 12$	68	$+35$	0.62
6342	17 18 .2	$-19\ 32$	5	$+\ 9$	1.35
6352	17 21 .6	$-48\ 26$	342	$-\ 7$	1.03
6355	17 20 .9	$-26\ 19$	359	$+\ 5$	1.58

EXERCISE III

DETERMINATION OF THE VERTEX OF
THE HYADES CLUSTER

1. The Hyades cluster in the constellation of Taurus is the best-known and most investigated moving cluster. The proper motions of the cluster members show a very conspicuous convergent character. Moreover they are of the same order of magnitude (Figure III.1). This has long ago been explained by assuming that all the cluster members are moving with equal velocities. The point on the celestial sphere towards which the total proper motions are directed is called the vertex of the moving cluster (or its convergent point).

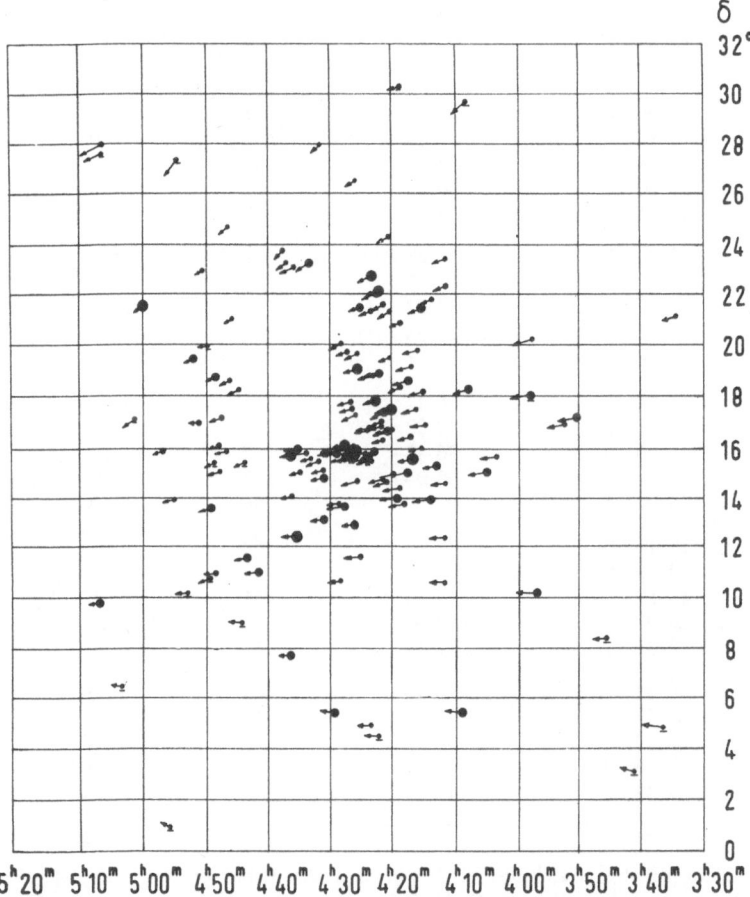

Fig. III.1.   From van Bueren [4] by kind permission of the author.

Table II.1 (continued)

NGC	α	δ	l	b	$B - V$
6356	17h20m.7	− 17°46′	7°	+ 10°	1.14
6362	17 26 .6	− 67 01	326	− 17	0.90
6366	17 25 .1	− 05 02	18	+ 16	1.60:
6388	17 32 .6	− 44 43	345	− 7	1.16
6397	17 36 .8	− 53 39	339	− 12	0.76
6401	17 35 .6	− 23 53	3	+ 4	1.32:
6402	17 35 .0	− 03 15	21	+ 14	1.28
6426	17 42 .4	+ 03 12	28	+ 16	0.98
6440	17 45 .9	− 20 21	7	+ 3	2.00
6441	17 46 .8	− 37 02	353	− 5	1.27
6453	17 48 .0	− 34 37	355	− 4	1.18:
6517	17 59 .1	− 08 57	19	+ 6	1.84
6522	18 00 .4	− 30 02	1	− 4	1.19
6528	18 01 .6	− 30 04	0	− 5	1.43
6535	18 01 .3	− 00 18	27	+ 10	0.97
6539	18 02 .1	− 07 35	21	+ 6	1.87
6541	18 04 .4	− 43 44	349	− 11	0.76
6544	18 04 .3	− 25 01	6	− 2	1.43
6553	18 06 .3	− 25 56	5	− 3	1.62
6569	18 10 .4	− 31 50	0	− 7	1.28
6584	18 14 .6	− 52 14	343	− 17	0.79
6624	18 20 .5	− 30 23	2	− 8	1.11
6626	18 21 .5	− 24 54	7	− 6	1.09
6637	18 28 .1	− 32 23	1	− 11	0.99
6638	18 27 .9	− 25 32	8	− 8	1.12
6652	18 32 .5	− 33 02	1	− 12	0.89
6656	18 33 .3	− 23 58	9	− 8	1.00
6681	18 40 .0	− 32 21	2	− 13	0.71
6712	18 50 .3	− 08 47	27	− 5	1.16
6715	18 52 .0	− 30 32	5	− 15	0.83
6723	18 56. 2	− 36 42	0	− 18	0.74
6752	19 06 .4	− 60 04	337	− 26	0.65
6760	19 08 .6	+ 00 57	36	− 4	1.73
6779	19 14 .6	+ 30 05	62	+ 9	0.88
6809	19 36 .9	− 31 03	9	− 24	0.69
6838	19 51 .5	+ 18 39	56	− 5	1.15
6864	20 03 .2	− 22 04	20	− 26	0.85
6934	20 31 .7	+ 07 14	52	− 19	0.76
6981	20 50 .7	− 12 44	35	− 33	0.73
7006	20 59 .1	+ 16 00	64	− 19	0.78
7078	21 27 .6	+ 11 57	65	− 27	0.69
7089	21 30 .9	− 01 03	54	− 36	0.68
7099	21 37 .5	− 23 25	27	− 47	0.59

The equatorial coordinates of globular clusters given in this Table refer to 1950.0. The (new) galactic coordinates have been taken from H. C. Arp's list (Arp, 1965). Color indexes are from S. van den Bergh's Table IV (Van den Bergh, 1967). A colon (:) indicates an uncertain value.

It is well known, from General Astronomy courses, that if the position of the vertex is known, and if the radial velocities of at least some of the cluster members have been measured, one can determine the space velocity of the cluster as well as its distance from the Sun. In fact, denote by λ the spherical distance of a cluster member from the vertex, by μ its total proper motion (in seconds of arc per year) and by ϱ its radial velocity (in kilometres per second). Then, as shown in any text-book of General Astronomy, the stars' distance from the Sun, r, in parsecs, can be found by

$$r = \frac{V \sin \lambda}{4.738 \, \mu} \tag{III.1}$$

where V is the so-called space velocity of the cluster (or, to be more precise, the absolute value of the space velocity vector). This later quantity can be found from the equation

$$V = \varrho \sec \lambda. \tag{III.2}$$

The distances of the moving cluster members are by far the best stellar distances available. The corresponding parallaxes are called cluster parallaxes. Now, starting from distances and apparent magnitudes one can easily find the absolute magnitudes. Moreover, by measuring the color indexes it is possible to derive very accurate color-magnitude diagrams. In particular, the color-magnitude diagram of the Hyades cluster has been of prime importance in establishing the so-called Zero Age Main Sequence as well as in calibrating luminosity criteria, which permit the determination of the absolute magnitudes of the stars from observable features in their spectra. Therefore the determination of the vertex of the Hyades cluster must be considered as an essential part of a complex problem which has very wide implications.

In this exercise we propose that the reader should find the equatorial coordinates of the vertex of the Hyades cluster and derive its space velocity.

2. Our problem can be stated as follows:

It is assumed that the cluster members move with equal space velocities. The equatorial coordinates and the proper motion components of the cluster stars are known. Determine the equatorial coordinates of the vertex.

It should be stressed that by space velocity we shall always understand the space velocity vector, although this term is often used only for its absolute value (in km/s).

Let us now derive the formulae which solve our problem.

Take a coordinate system having its origin in the Sun, the axes being oriented as follows:

the x-axis towards the point $\alpha = 0$ h, $\delta = 0°$
the y-axis towards the point $\alpha = 6$ h, $\delta = 0°$
the z-axis towards the North Celestial Pole,

of a definite epoch. Let $\mathbf{x}^0$, $\mathbf{y}^0$, and $\mathbf{z}^0$, be the unit vectors corresponding to these three axes respectively. Denote by S_i a member of the moving cluster. Taking the Sun as the center draw a sphere through the star S_i. Trace on this sphere the circles of constant right ascension and the circles of constant declination which correspond to S_i (Figures

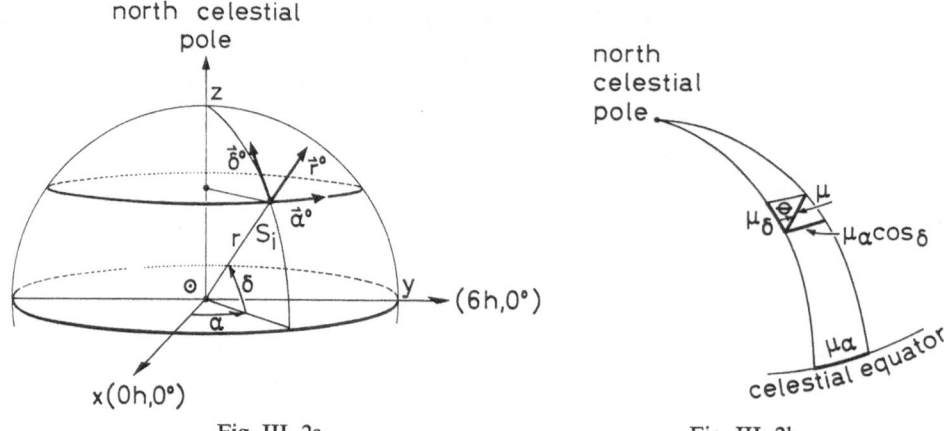

Fig. III. 2a.   Fig. III. 2b.

III.2). Then, taking S_i as the origin, construct a trihedral consisting of three mutually perpendicular unit vectors defined as follows:

The unit vector $\boldsymbol{\alpha}^0$ tangent to the circle of constant declination and pointing in the direction of growing right ascensions;

The unit vector $\boldsymbol{\delta}^0$ tangent to the circle of constant right ascension and pointing towards the North Celestial Pole;

The unit vector $\mathbf{r}^0$ lying on the radius vector, which joins the Sun to the star, and pointing from the sphere.

This trihedral is sometimes called the astronomical trihedral.

Denote by α and δ the right ascension and the declination of the star. Then it can be proved that

$$\begin{aligned}
\boldsymbol{\alpha}^0 &= -\sin\alpha\cdot\mathbf{x}^0 && + \cos\alpha\cdot\mathbf{y}^0 \\
\boldsymbol{\delta}^0 &= -\sin\delta\cos\alpha\cdot\mathbf{x}^0 - \sin\delta\sin\alpha\cdot\mathbf{y}^0 + \cos\delta\cdot\mathbf{z}^0 \\
\mathbf{r}^0 &= +\cos\delta\sin\alpha\cdot\mathbf{x}^0 + \cos\delta\sin\alpha\cdot\mathbf{y}^0 + \sin\delta\cdot\mathbf{z}^0 .
\end{aligned} \qquad \text{(III.3)}$$

The proof is left to the reader (See Problem III.1 at the end of this exercise).

Strictly speaking all the quantities proper to the cluster member S_i and changing from one star to another, that is in fact all of them, except only the space velocity, should be denoted by the subscript i. In order to avoid cumbersome notations this subscript will however be introduced only in the final equation.

Let now $\mathbf{V}$ be the vector of the space velocity of S_i relative to the Sun. With the origin in the Sun take a vector having the same intensity as, and parallel to, the space velocity $\mathbf{V}$. Denote its components along the x, y, and z, axes by X, Y, and Z, respectively. Let A and D be the right ascension and the declination of the point towards which it is directed. Then A and D are the equatorial coordinates of the vertex. We have the well-known relations:

$$X = V\cdot\cos D\cdot\cos A , \quad Y = V\cdot\cos D\cdot\sin A , \quad Z = V\cdot\sin D . \qquad \text{(III.4)}$$

We shall always measure the components of V in km/s.

Put $x = X/Z$ and $y = Y/Z$. Then

$$\cot A = x/y \quad \text{and} \quad \operatorname{tg} D = (x^2 + y^2)^{1/2}. \tag{III.5}$$

Take now the scalar product of

$$\mathbf{V} = X \cdot \mathbf{x}^0 + Y \cdot \mathbf{y}^0 + Z \cdot \mathbf{z}^0 \tag{III.6}$$

with $\boldsymbol{\alpha}^0$, $\boldsymbol{\delta}^0$, and $\mathbf{r}^0$. This gives the following three equations

$$\begin{aligned}
\mathbf{V} \cdot \boldsymbol{\alpha}^0 &= -X \sin\alpha && + Y \cos\alpha && = V_\alpha, \\
\mathbf{V} \cdot \boldsymbol{\delta}^0 &= -X \sin\delta \cos\alpha && - Y \sin\delta \sin\alpha + Z \cos\delta && = V_\delta, \\
\mathbf{V} \cdot \mathbf{r}^0 &= +X \cos\delta \cos\alpha && + Y \cos\delta \sin\alpha + Z \sin\delta && = \varrho.
\end{aligned} \tag{III.7}$$

V_α, V_δ, and ϱ, are the components of V along the axes of the astronomical trihedral. Let us now interpret these components in terms of observable quantities.

The component $\mathbf{V} \cdot \mathbf{r}^0 = \varrho$ affects only the star's distance r from the Sun. The quantity ϱ is the radial velocity of the star relative to the Sun. For obvious reason ϱ is taken positive if the star is receding, negative if it is approaching.

Both the components V_α and V_δ are in the plane tangent to the sphere and drawn through the star. Therefore they affect only the direction of $\mathbf{r}^0$, that is the equatorial coordinates of the star. In particular V_α affects only the star's right ascension, whereas V_δ affects only its declination.

Let the question of the units remain open for a while, and denote by μ_δ the change of the star's declination due to V_δ. Obviously

$$V_\delta = k \cdot r \cdot \mu_\delta, \tag{III.8}$$

the value of the coefficient k depending on the units chosen. When the distance r is measured in parsecs and the change in declination expressed in seconds of arc per tropical year ($''/a$), this coefficient has the value $k = 4.738$. The quantity μ_δ is the proper motion of the star in declination.

An analogous relation holds for V_α and the corresponding variation μ_α of the right ascension of the star. Note however that V_α is related to the arc of the circle of constant declination which has a radius equal to $r \cdot \cos\delta$. Accordingly

$$V_\alpha = 4.738 \cdot r \cdot \mu_\alpha \cdot \cos\delta, \tag{III.9}$$

where μ_α is expressed in $''/a$. Often μ_α is given in seconds of time per year, as the right ascension itself is expressed in time units. In this case it should be converted into $''/a$ by multiplying it by 15. The quantity μ_α is the star's proper motion in right ascension. The quantities μ_α and μ_δ may be called the equatorial components of the proper motion. If the units are suitably chosen, they may be defined as the angles subtended by V_α and V_δ respectively.

The component of $\mathbf{V}$ in the tangential plane is the tangential velocity V_t. Obviously we have

$$V_t = V_\alpha^2 + V_\delta^2. \tag{III.10}$$

The angle subtended by V_t, the units being chosen as above, is the total proper motion μ of the star. Evidently

$$V_t = 4.738 \cdot r \cdot \mu. \tag{III.11}$$

By combining this with the Equations (III.8), (III.9) and (III.10) we get

$$\mu^2 = \mu_\alpha^2 \cos^2 \delta + \mu_\delta^2 \,. \tag{III.10a}$$

In our preceding discussion the various μ's have been defined by the variations of the unit vector pointing from the Sun to the star. Adopting the conventional standpoint and using the concept of the celestial sphere (distinct from the sphere of radius r drawn through the star), the same quantities can be defined as displacements on the sphere or as elements of arcs of circles traced on it. The relations between the different μ's can then readily be seen from Figure III.2b. In particular the direction of the total proper motion μ, is defined by the position angle θ, given by

$$\operatorname{tg} \theta = \mu_\alpha \cdot \cos \delta / \mu_\delta \,. \tag{III.12}$$

Notice that, as μ, $\mu_\alpha \cos \delta$, and μ_δ, are very small quantities we can consider them as elementary vectors in the plane tangent to the sphere.

Now substitute in the first two Equations (III.7) for V_α and V_δ their values as given by (III.8) and (III.9). Eliminate the unknown distance r by multiplying the first equation by μ_δ, and the second one by $-\mu_\alpha \cos \delta$, and add. This gives the following final, equation

$$a_i x + b_i y = c_i \tag{III.13}$$

where the coefficients

$$\begin{aligned}
a_i &= \mu_\alpha \cos \delta \cos \alpha \sin \delta - \mu_\delta \sin \alpha \\
b_i &= \mu_\alpha \cos \delta \sin \alpha \sin \delta + \mu_\delta \cos \alpha \\
c_i &= \mu_\alpha \cos^2 \delta \,,
\end{aligned} \tag{III.14}$$

have been denoted by the subscript i, as their value obviously depends on the star chosen. Notice however that according to our basic assumption concerning the velocities of the stars, the quantities x and y will be the same for all members of the moving cluster.

For each cluster member for which the proper motion components are known and which we intend to use for the derivation of the vertex we can write down an equation of this form, and then solve the system by making use of the Principle of Least Squares. (For a short account on least squares solutions the reader is referred to Appendix I, p. 132.) In other words the Equations (III.13) are the equations of condition of our problem. The normal equations will be:

$$\begin{aligned}
[aa] \, x + [ab] \, y &= [ac] \\
[ba] \, x + [bb] \, y &= [bc]
\end{aligned} \tag{III.15}$$

where as usual

$$\begin{aligned}
[aa] &= \sum_{i=1}^{n} a_i^2, \qquad [ab] = \sum_{i=1}^{n} a_i b_i, \qquad [ac] = \sum_{i=1}^{n} a_i c_i, \\
[ba] &= [ab] \qquad\quad [bb] = \sum_{i=1}^{n} b_i^2, \qquad [bc] = \sum_{i=1}^{n} b_i c_i \,.
\end{aligned} \tag{III.16}$$

Having found x and y we can compute the equatorial coordinates A and D of the vertex by formulae (III.5).

In order to determine the absolute value of the space velocity consider the third of the Equations (III.7) and remember that, according to the definition of the scalar product

$$\varrho = V \cos \lambda \qquad\qquad (III.17)$$

where V is the absolute value of the space velocity, and where λ is the angle between $\mathbf{V}$ and $\mathbf{r}^0$. This angle is obviously equal to the spherical distance of the star S_i from the vertex. It can be found from (see Figure III.3)

$$\cos \lambda = \cos D \cos \delta \cos (A - \alpha) + \sin D \sin \delta \qquad\qquad (III.18)$$

an equation which follows from (III.17) and the third of the Equations (III.7), by substituting for X, Y, and Z, their values given by (III.4). It can also be derived by applying the cosine theorem of spherical trigonometry to the triangle North Celestial Pole – Star – Vertex (see Figure III.3).

Equation (III.17) is the equation of condition for the determination of the absolute

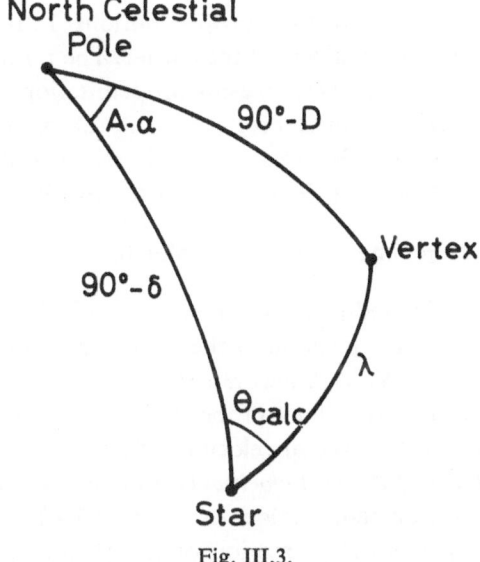

Fig. III.3.

value of the space velocity. The corresponding normal equation is

$$V [\cos^2 \lambda] = [\varrho \cos \lambda] . \qquad\qquad (III.19)$$

The Equations (III.15) and (III.19) solve the problem completely.

The method of solution shown above is due to C. V. L. Charlier [1]. A simpler and more direct derivation of the equation of condition (III.13) is given in the solution to Problem III.2 at the end of this Exercise. For an account of other methods the reader is referred to [2].

Having computed the equatorial coordinates of the vertex we shall compare the directions of the proper motions according to our solution, to the observed directions.

The directions of the proper motion as given by observation, which we shall denote by θ_{obs}, can be found from Equation (12) which we will use in the form

$$\cot \theta_{obs} = \mu_\delta / \mu_\alpha \cos \delta .$$

On the other hand, notice that according to our solution the direction of the proper motion will be that of the arc of the great circle joining the star to the vertex. Therefore the calculated direction, which we shall denote by θ_{calc}, is given by the angle on the star, in the spherical triangle Star – North Celestial Pole – Vertex (see Figure III.3). Applying the cosine theorem of spherical trigonometry to this triangle and solving for θ_{calc} we easily find

$$\cos \theta_{calc} = (\sin D - \cos \lambda \sin \delta)/\sin \lambda \cos \delta . \tag{III.19}$$

3. Let us now quote some sources as well as some results.

The first determination of the motion of the Hyades cluster was made in 1908 by L. Boss. For the proper motion he used the values from his *Preliminary General Catalogue* (PGC). On the ground of their proper motions, L. Boss identified 39 stars in the region of the Hyades as members of the cluster. The position of the vertex was computed (by a method of trial and error) using proper motions of only 13 stars, these being considered as being of highest precision. The (absolute value of the) space velocity was determined using only three radial velocities (measured by F. Küstner). L. Boss's result (the coordinates of the vertex reduced to 1950.0) is:

$$A = 6 \text{ h } 11 \text{ m} \pm 6 \text{ m}, \qquad D = + 6°55' \pm 0°3, \qquad V = + 45.6 \text{ km/s} .$$

The errors given are estimated probable errors.

The last determination of the position of the vertex based on catalogue values for the proper motions is due to W. M. Smart [3]. He used data from the *Albany General Catalogue* (GC). On the ground of their proper motions, Smart identified 72 stars as members of the Hyades cluster. Radial velocities of 35 of these stars appear in J. H. Moore's *General Catalogue of Radial Velocities* (*Lick Obs. Publ.* **18**, 1932), at that time the fundamental reference on radial velocities. Using Charlier's method, Smart deduced the following elements of the motion of the Hyades cluster (epoch 1950.0)

$$A = 6 \text{ h } 07 \text{ m} \pm 2.6 \text{ m} \qquad D = + 8°09' \pm 16', \qquad V = 42.5 \text{ km/s} \pm 0.30 \text{ km/s} .$$

The reader is referred to Smart's paper for its very clear exposition as well as for bibliographic references on earlier determinations made by different investigators from the time of L. Boss on.

In a thorough study of the Hyades cluster H. G. van Bueren [4] made use of corrected values for the proper motion as well as of more recent and better values of the radial velocities. Table 2a of his paper contains data for 132 stars considered as certain members of the cluster. The numbers by which the stars are listed in this and other tables in van Bueren's paper are now currently used to designate Hyades cluster members.

The solution as found by van Bueren is:

$$A = 6 \text{ h } 18.5 \text{ m} \pm 2.1 \text{ m} \qquad D = + 7°29' \pm 11' \qquad V = 43.95 \text{ km/s} \pm 0.60 \text{ km/s}.$$

(the errors given are mean errors).

The most recent study of the motion of the Hyades cluster is due to P. A. Wayman et al. [5]. Of the 132 Hyades cluster members listed in van Bueren's Table 2a, the data collected in their paper indicate that two (No. 98 and 125) are probably not cluster members. On the other hand, investigation of 99 stars in a nearby area indicates that, on the ground of their radial velocity, about 14 of these stars probably are cluster members. For the position of the vertex these authors find

$$A = 6 \text{ h } 16.5 \text{ m} \pm 1.3 \text{ m} \qquad D = + 7°41' \pm 8' \qquad (1950.0).$$

The values quoted represent the weighted mean of four independent solutions, each based on the proper motions of thirty to fifty stars, taken from different sources and corrected. According to Wayman et al., the absolute value of the space velocity is

$$V = 43.25 \text{ km/s} \pm 0.13 \text{ km/s}$$

(here and above probable errors are given).

Table VI of the paper cited contains the best proper motion data for twenty Hyades stars, at present available. (The heading μ_α in column five of this table stands for $\mu_\alpha \cos\delta$.)

It is unfortunately impossible to discuss here the very important question concerning the correction of proper motion data contained in various fundamental catalogues. In any case the reader can use, for the proper motions, data listed in van Bueren's Table 2a (the numbers of stars with most accurate proper motions are printed in italics) or values quoted in Table VI in the paper by Wayman et al. In order to identify stars in this list (HD numbers are given) use e.g. Bečvář's *Atlas Coeli II – Katalog* [6] or the *Yale Bright Star Catalogue* [7]. For the radial velocities the fundamental reference is R. E. Wilson's *General Catalogue of Radial Velocities* [8]. Wayman et al. give also some more recent and accurate data.

4. Now that all the necessary formulae have been deduced, and the principal sources of observational data quoted, it is possible to undertake the determination of the motion of the Hyades cluster.

Owing to the large amount of computational work involved, we shall be obliged to use a rather limited number of stars. But even so, and thanks to the very pronounced convergence of the total proper motion, we can expect to get a good result, if the stars are properly chosen.

The determination of the equatorial coordinates of the vertex and of the space velocity, to be made in this exercise, will be based on ten stars.

If, as we will do, the proper motion data are taken from van Bueren's list (*loc. cit.*, Table 2a), the choice shall be made among stars with their numbers printed in italics.

Radial velocity data can be found in Wilson's General Catalogue of Radial Velocities. Note, however, that the values given are of different accuracy. Their quality is expressed, in order of decreasing accuracy, by the letters a, b, c, d and e. Radial velocities of quality a, b, are to be preferred, and those of quality c should be used only if unavoidable.

The stars chosen shall form a group covering the whole cluster. Once the stars have been selected according to the accuracy with which their motion is known, their position within the cluster can be found using the chart given by van Bueren (Figure 4 of his paper).

This part of the work has already been done. Table III.1 (p. 46) contains all the necessary data for the ten stars selected (including the trigonometric functions of the equatorial coordinates). It is proposed to the reader that he should

(1) make a solution using all ten stars from Table III.1;

(2) compare the observed values of the position angle of the total proper motions with $\theta_{i(\text{calc})}$ and discuss the differences found;

(3) determine the space velocity of the cluster.

The reader is referred to Appendix I for the derivation of normal equations by the principle of least squares.

It has been objected (see, e.g. [2]) that in the present case a least-square solution can yield only approximate (but in fact quite good) values of A and D. The corrections ΔA and ΔD, which one has to apply to A and D, respectively, in order to get the correct values of the coordinates of the vertex, can be found as follows.

Consider the first two of the Equations (III.7)

$$V_\alpha = - X \sin \alpha \qquad + Y \cos \alpha$$
$$V_\delta = - X \sin \delta \cos \alpha - Y \sin \delta \sin \alpha + Z \cos \delta$$

and notice that, according to definition, the position angle of the total proper motion can also be found from

$$\cot \theta = \frac{V_\delta}{V_\alpha} = \frac{X \sin \delta \cos \alpha + Y \sin \delta \sin \alpha - Z \cos \delta}{X \sin \alpha - Y \cos \alpha}. \tag{III.20}$$

By making use of the Equations (III.4) this gives, after some simple transformations,

$$\cot \theta = \cos \delta \tan D \csc (A - \alpha) - \sin \delta \cot (A - \alpha). \tag{III.21}$$

Now, if ΔA is a small variation in A, and ΔD a small variation in D, and if, moreover, terms of an order higher than the first in ΔA and ΔD can be neglected, then the corresponding variation $\Delta \theta$ can be found by simply differentiating (III.21). This gives

$$\Delta \theta = H \cdot \Delta A + K \cdot \Delta D, \tag{III.22}$$

where

$$H = \frac{\sin^2 \theta}{\sin^2 (A - \alpha)} (\cos \delta \tan D \cos (A - \alpha) - \sin \delta),$$

and

$$K = \frac{- \sin^2 \theta}{\sin^2 (A - \alpha)} \cos \delta \sec^2 D \sin (A - \alpha). \tag{III.23}$$

If now we already have obtained good approximate values for A and D, then (III.22) can be considered as an equation of condition permitting the corrections ΔA and ΔD to be derived from $\Delta \theta$. For $\Delta \theta$ we shall take the differences between the values θ_{obs} which follow from the observed proper motions and the values θ_{calc} according to our (approximate) solution, i.e. we shall put

$$\Delta \theta = \theta_{\text{obs}} - \theta_{\text{calc}} . \tag{III.24}$$

As an extension of the present exercise we propose that the more interested reader should determine the corrections ΔA and ΔD and derive the corrected values of the coordinates of the vertex, A^* and D^*, which will be given by:

$$\begin{aligned} A^* &= A + \Delta A, \\ D^* &= D + \Delta D. \end{aligned} \tag{III.25}$$

For a thorough discussion of the possible errors in the distances of the Hyades cluster stars as derived by the method explained above, the reader is referred to an article by P. A. Wayman [9].

Problem III.1 Formulae for the Unit Vectors of the Astronomical Trihedral

Prove formulae (III.3)

$$\begin{aligned}
\boldsymbol{\alpha}^0 &= - \qquad \sin\alpha_i \mathbf{x}^0 + \qquad \cos\alpha_i \mathbf{y}^0 \\
\boldsymbol{\delta}^0 &= -\sin\delta_i \cos\alpha_i \mathbf{x}^0 - \sin\delta_i \sin\alpha_i \mathbf{y}^0 + \cos\delta_i \mathbf{z}^0 \\
\mathbf{r}^0 &= +\cos\delta_i \cos\alpha_i \mathbf{x}^0 + \cos\delta_i \sin\alpha_i \mathbf{y}^0 + \sin\delta_i \mathbf{z}^0 .
\end{aligned}$$

Solution

(a) Any vector $\mathbf{v}$ can be represented in the form:

$$\mathbf{v} = (\mathbf{v}\cdot\mathbf{x}^0)\,\mathbf{x}^0 + (\mathbf{v}\cdot\mathbf{y}^0)\,\mathbf{y}^0 + (\mathbf{v}\cdot\mathbf{z}^0)\,\mathbf{z}^0 . \qquad (\text{P.III.1})$$

In order to prove formulae ('III.3) we shall therefore compute the nine scalar products

$$\begin{aligned}
&\boldsymbol{\alpha}^0\cdot\mathbf{x}^0 \quad \boldsymbol{\alpha}^0\cdot\mathbf{y}^0 \quad \boldsymbol{\alpha}^0\cdot\mathbf{z}^0 \\
&\boldsymbol{\delta}^0\cdot\mathbf{x}^0 \quad \boldsymbol{\delta}^0\cdot\mathbf{y}^0 \quad \boldsymbol{\delta}^0\cdot\mathbf{z}^0 \\
&\mathbf{r}^0\cdot\mathbf{x}^0 \quad \mathbf{r}^0\cdot\mathbf{y}^0 \quad \mathbf{r}^0\cdot\mathbf{z}^0
\end{aligned} \qquad (\text{P.III.2})$$

or, what amounts to the same, we shall find the orthogonal projections of each of the three unit vectors $\boldsymbol{\alpha}^0$, $\boldsymbol{\delta}^0$ and $\mathbf{r}^0$ onto the axes x, y, and z.

By definition $\boldsymbol{\alpha}^0$ is in a plane parallel to the xOy plane and, accordingly, perpendicular to $\mathbf{z}^0$. Therefore

$$\boldsymbol{\alpha}^0\cdot\mathbf{z}^0 = 0 . \qquad (\text{P.III.3})$$

In order to find the two other components, project $\boldsymbol{\alpha}^0$ into the xOy plane and then on the x and y axes. It is easily found that (Figure III.4)

$$\boldsymbol{\alpha}^0\cdot\mathbf{x}^0 = -\sin\alpha \qquad \boldsymbol{\alpha}^0\cdot\mathbf{y}^0 = +\cos\alpha . \qquad (\text{P.III.4})$$

Therefore

$$\boldsymbol{\alpha}^0 = -\sin\alpha\,\mathbf{x}^0 + \cos\alpha\,\mathbf{y}^0 \qquad (\text{P.III.5})$$

which is the first of formulae (III.3).

By definition $\boldsymbol{\delta}^0$ is in a plane drawn through the z-axis and making an angle α with the xOz plane. From Figure III.4 one immediately finds

$$\boldsymbol{\delta}^0\cdot\mathbf{z}^0 = \cos\delta . \qquad (\text{P.III.6})$$

In order to find the other two components, project $\boldsymbol{\delta}^0$ in the xOy plane and there-

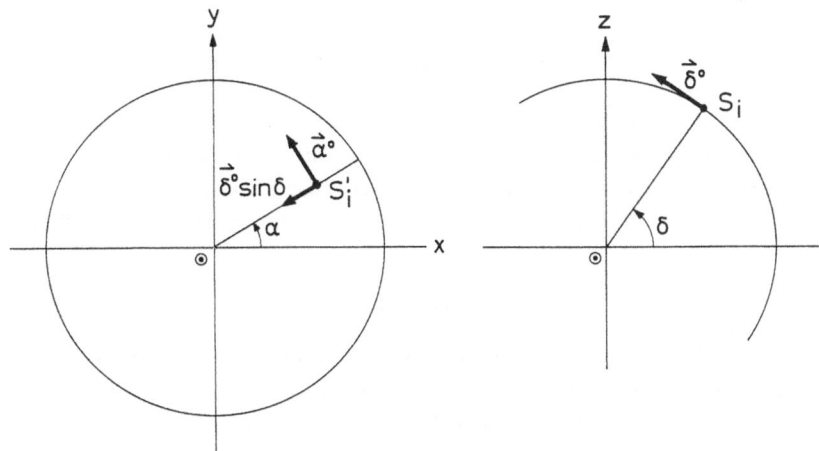

Fig. III.4.

after onto the x and y axes respectively. This gives (Figure III.4)

$$\boldsymbol{\delta}^0 \cdot \mathbf{x}^0 = -\sin\delta\cos\alpha \qquad \boldsymbol{\delta}^0 \cdot \mathbf{y}^0 = -\sin\delta\sin\alpha.$$ (P.III.7)

Accordingly

$$\boldsymbol{\delta}^0 = (\boldsymbol{\delta}^0 \cdot \mathbf{x}^0)\,\mathbf{x}^0 + (\boldsymbol{\delta}^0 \cdot \mathbf{y}^0)\,\mathbf{y}^0 + (\boldsymbol{\delta}^0 \cdot \mathbf{z}^0)\,\mathbf{z}^0$$
$$= -\sin\alpha\cos\delta\ \mathbf{x}^0 - \sin\delta\sin\alpha\ \mathbf{y}^0 + \cos\delta\ \mathbf{z}^0$$ (P.III.8)

which is the second of formulae (III.3)

The formula for $\mathbf{r}^0$ is well-known from elementary vector algebra.

(b) Another proof is the following one.

Start from the well-known formula for $\mathbf{r}^0$:

$$\mathbf{r}^0 = \cos\delta\cos\alpha\ \mathbf{x}^0 + \cos\delta\sin\alpha\ \mathbf{y}^0 + \sin\delta\ \mathbf{z}^0.$$ (P.III.9)

Remember that, as the unit vectors $\boldsymbol{\alpha}^0$, $\boldsymbol{\delta}^0$, and $\mathbf{r}^0$ form an orthogonal righthanded trihedral, one has

$$\boldsymbol{\alpha}^0 \times \boldsymbol{\delta}^0 = \mathbf{r}^0, \qquad \boldsymbol{\delta}^0 \times \mathbf{r}^0 = \boldsymbol{\alpha}^0, \qquad \mathbf{r}^0 \times \boldsymbol{\alpha}^0 = \boldsymbol{\delta}^0.$$ (P.III.10)

Therefore, if any two of the three unit vectors are known, the third one can be determined with the aid of the corresponding formula of this system.

Now from Figure III.2 and Figure III.4, and bearing in mind that $\boldsymbol{\alpha}^0$ is perpendicular to both $\mathbf{z}^0$ and $\mathbf{r}^0$, one easily finds that

$$\mathbf{z}^0 \times \mathbf{r}^0 = \boldsymbol{\alpha}^0 \sin(\mathbf{z}^0, \mathbf{r}^0) = \boldsymbol{\alpha}^0 \cos\delta.$$ (P.III.11)

Substitute, for $\mathbf{r}^0$, the expression given above, and use for the cross products of the unit vectors $\mathbf{x}^0$, $\mathbf{y}^0$, and $\mathbf{z}^0$, the relations analogous to those given for $\boldsymbol{\alpha}^0$, $\boldsymbol{\delta}^0$, and $\mathbf{r}^0$. The expression for $\boldsymbol{\alpha}^0$ follows immediately.

Finally use the third of the Equations (P.III.10) to compute $\boldsymbol{\delta}^0$.

Problem III.2.   An Alternative Derivation of the Fundamental Formula of Charlier's Method and Its Geometric Interpretation

Let $\mathbf{r}^0$ and $\mathbf{v}^0$ respectively be the unit vectors pointing from the Sun towards a member of a moving cluster and towards the vertex. Consider the total proper motion μ as a vector. Show that the equation of condition (III.13) for the determination of the equatorial coordinates A and D of the vertex is, in fact, the condition of coplanarity of these three vectors.

Solution

The coplanarity of the three vectors $\mathbf{r}^0$, $\mathbf{v}^0$, and μ is geometrically evident (Figure III.5).

The total proper motion considered as a vector in the plane tangent to the celestial sphere and drawn through the star, can be decomposed into two components along

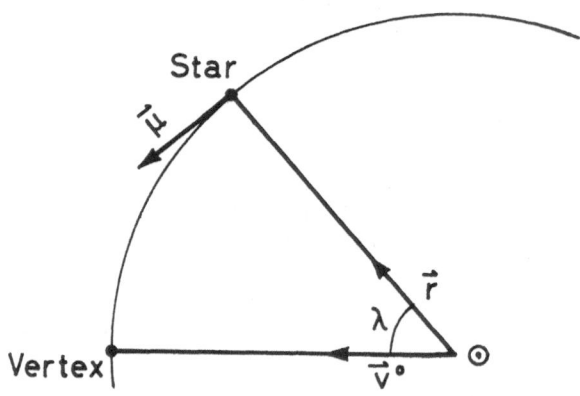

Fig. III.5.

the axes defined by α^0 and δ^0 as follows:

$$\mu = \xi \alpha^0 + \eta \delta^0 ; \quad \xi = \mu_\alpha \cos \delta, \quad \eta = \mu_\delta. \tag{P.III.12}$$

The condition of coplanarity of the three vectors $\mathbf{r}^0$, $\mathbf{v}^0$, and μ is

$$\mathbf{r}^0 \cdot (\mathbf{v}^0 \times \mu) = \mathbf{v}^0 \cdot (\mu \times \mathbf{r}^0) = 0. \tag{P.III.13}$$

As $\mathbf{r}^0$ is a unit vector perpendicular to the plane containing μ, the cross-product

$$\mu \times \mathbf{r}^0 = - (\mathbf{r}^0 \times \mu) \tag{P.III.14}$$

is simply obtained by rotating μ clockwise by 90°. Therefore:

$$\mu \times \mathbf{r}^0 = + \eta \alpha^0 - \xi \delta^0 . \tag{P.III.15}$$

Denote by A and D respectively the right ascension and the declination of the vertex.

Then $\mathbf{v}^0$ can be put into the form:

$$\mathbf{v}^0 = L\mathbf{x}^0 + M\mathbf{y}^0 + N\mathbf{z}^0 \qquad \text{(P.III.16)}$$

where

$$L = \cos D \cos A, \qquad M = \cos D \sin A, \qquad N = \sin D. \qquad \text{(P.III.17)}$$

The condition for coplanarity then becomes:

$$\mathbf{v}^0 \cdot (\boldsymbol{\mu} \times \mathbf{r}^0) = a \cdot L + b \cdot M - c \cdot N = 0 \qquad \text{(P.III.18)}$$

the quantities a, b, and c being identical to those given by Equation (III.14).

Divide the foregoing equation by N and note that

$$L/N = X/Z = x, \qquad M/N = Y/Z = y, \qquad \text{(P.III.19)}$$

the quantities x, y, and X, Y, and Z having the same meaning as before.

The condition of coplanarity, written in the form

$$ax + by = c \qquad \text{(P.III.20)}$$

is therefore identical to the equation of condition (III.13).

References

[1] Charlier, C. V. L.: 1926, *The Motion and the Distribution of the Stars*, University of California Press, Berkeley, Chapter IV.
[2] Smart, W. M.: 1968, *Stellar Kinematics*, Longmans, Green and Co., London, 1968, Chapter 7.
[3] Smart, W. M.: 1939, *Monthly Notices Roy. Astron. Soc.* **99**, 168.
[4] Van Bueren, H. G.: 1952, *Bull. Astron. Inst. Neth.* **XI**, 385.
[5] Wayman, P. A., Symms, L. S. T., and Blackwell, K. C.: 1965, *Royal Obs. Bull.* No. 98.
[6] Bečvář, A.: 1960, *Atlas Coeli II, Katalog 1950.0*, Czechoslovak Academy of Sciences, Praha.
[7] Hoffleit, D.: 1964, *Yale Catalogue of Bright Stars*, Yale University Observatory, New Haven.
[8] Wilson, R. E.: 1963, *General Catalogue of Stellar Radial Velocities*, Carnegie Institution, Washington D.C.
[9] Wayman, P. A.: 1967, *Publ. Astron. Soc. Pacific*, **79**, 156.

TABLE III.1ᵃ

The equatorial coordinates, proper motion components, and radial velocities of some Hyades stars

vB	GC	α	δ	$\sin\alpha$	$\cos\alpha$	$\sin\delta$	$\cos\delta$	$\mu_\alpha\cos\delta$	μ_δ	μ	ϱ
6	4677	3ʰ50ᵐ18ˢ	+17°10'47"	+0.8441	+0.5362	+0.2954	+0.9554	+149 ±2	−28 ±2	151	+35. c
14	5042	4 08 40	5 23 40	0.8843	0.4669	0.0940	0.9956	152 2	+10 2	152	36.6b
20	5137	4 12 56	15 16 38	0.8928	0.4504	0.2635	0.9647	121 2	−29 2	124	36.9b
41	5304	4 20 03	17 25 37	0.9064	0.4224	0.2995	0.9541	112 1	−31 1	115	38.4a
60	5370	4 23 19	22 42 07	0.9124	0.4095	0.3859	0.9225	110 2	−47 1	119	35.1b
74	5443	4 26 02	12 56 18	0.9171	0.3986	0.2239	0.9746	113 3	−12 2	113	33.4b
107	5665	4 36 24	7 46 24	0.9342	0.3567	0.1353	0.9908	92 4	+02 3	92	36.3b
108	5666	4 36 25	15 49 14	0.9342	0.3567	0.2726	0.9621	86 2	−17 2	88	36.9b
123	5907	4 48 27	18 45 23	0.9516	0.3072	0.3216	0.9469	83 2	−34 2	89	38.5b
130	6300	5 06 35	9 46 01	0.9730	0.2310	0.1696	0.9855	67 3	−03 2	67	37.2b

ᵃ The first two columns contain the number of the star in van Bueren's list [4] and in the General Catalogue, respectively. The equatorial coordinates refer to the equinox 1950.0. The proper motion components are those given by van Bueren. The radial velocities have been taken from R. E. Wilson's catalogue [8]. The units are 0.001″/a for the proper motions and km/s for the radial velocities.

TABLE III.2[a]

Computation of the coefficients a, b, and c

vB	$\xi \sin\delta$	$\xi \sin\delta \cos\alpha$	$-\eta \sin\alpha$	a	$\xi \sin\delta \sin\alpha$	$+\eta \cos\alpha$	b	$\xi \cos\delta = c$
6	+44.0	+23.6	+23.6	+47.2	+37.1	−15.0	+22.1	+142.4
14	+14.3	+ 6.7	− 8.8	− 2.1	+12.6	+ 4.7	+17.3	+151.3
20	+31.9	+14.4	+25.9	+40.3	+28.5	−13.1	+15.4	+116.7
41	+33.5	+14.2	+28.1	+42.3	+30.4	−13.1	+17.3	+106.9
60	+42.4	+17.4	+42.9	+60.3	+38.7	−19.2	+19.5	+101.5
74	+25.3	+10.1	+11.0	+21.1	+23.2	− 4.8	+18.4	+110.1
107	+12.4	+ 4.4	− 1.9	+ 2.5	+11.6	+ 0.7	+12.3	+ 91.2
108	+23.4	+ 8.3	+15.9	+24.2	+21.9	− 6.1	+15.8	+ 82.7
123	+26.7	+ 8.2	+32.4	+40.6	+25.4	−10.4	+15.0	+ 78.6
130	+11.4	+ 2.6	+ 2.9	+ 5.5	+11.1	− 0.7	+10.4	+ 66.0

[a] The coefficients a, b, and c, of the equations of condition (III.13) have been computed by the formulae (III.14). $\xi = \mu_\alpha \cos\delta$, $\eta = \mu_\delta$.

TABLE III.3[a]

The coefficients of the equations of condition

a	b	c	s
+ 47.2	+ 22.1	+ 142.4	+ 211.7
− 2.1	+ 17.3	+ 151.3	+ 166.5
+ 40.3	+ 15.4	+ 116.7	+ 172.4
+ 42.3	+ 17.3	+ 106.9	+ 166.5
+ 60.3	+ 19.5	+ 101.5	+ 181.3
+ 21.1	+ 18.4	+ 110.1	+ 149.6
+ 2.5	+ 12.3	+ 91.2	+ 106.0
+ 24.2	+ 15.8	+ 82.7	+ 122.7
+ 40.6	+ 15.0	+ 78.6	+ 134.2
+ 5.5	+ 10.4	+ 66.0	+ 81.9
$\Sigma a = +281.9$	$\Sigma b = +163.5$	$\Sigma c = +1047.4$	$\Sigma s = +1492.8$

[a] We have: $s_i = a_i + b_i + c_i$; $\Sigma_i s_i = \Sigma_i a_i + \Sigma_i b_i + \Sigma_i c_i$; see Appendix I, Equation (A.I.5).

TABLE III.4[a]

The computation of the coefficients of the first normal equation

aa	ab	ac	as
+ 2227.84	+1043.12	+ 6721.28	+ 9992.24
+ 4.41	− 36.33	− 317.73	− 349.65
+ 1624.09	+ 620.62	+ 4703.01	+ 6947.72
+ 1789.29	+ 731.79	+ 4521.87	+ 7042.95
+ 3636.09	+1175.85	+ 6120.45	+10932.39
+ 445.21	+ 388.24	+ 2323.11	+ 3156.56
+ 6.25	+ 30.75	+ 228.00	+ 265.00
+ 585.64	+ 382.36	+ 2001.34	+ 2969.34
+ 1648.36	+ 609.00	+ 3191.16	+ 5448.52
+ 30.25	+ 57.20	+ 363.00	+ 450.45
$[aa] = +11997.43$	$[ab] = +5002.60$	$[ac] = +29855.49$	$[as] = +46855.52$

[a] We have: $a_i a_i + a_i b_i + a_i c_i = a_i s_i$ and $[aa] + [ab] + [ac] = [as]$, see Equation (A.I.6).

TABLE III.5 [a]

The computation of the coefficients of the second normal equation

ba	bb	bc	bs
+ 1 043.12	+ 488.41	+ 3 147.04	+ 4 678.57
− 36.33	+ 299.29	+ 2 617.49	+ 2 880.45
+ 620.62	+ 237.16	+ 1 797.18	+ 2 654.96
+ 731.79	+ 299.29	+ 1 849.37	+ 2 880.45
+ 1 175.85	+ 380.25	+ 1 979.25	+ 3 535.35
+ 388.24	+ 338.56	+ 2 025.84	+ 2 752.64
+ 30.75	+ 151.29	+ 1 121.76	+ 1 303.80
+ 382.36	+ 249.64	+ 1 306.66	+ 1 938.66
+ 609.00	+ 225.00	+ 1 179.00	+ 2 013.00
+ 57.20	+ 108.16	+ 686.40	+ 851.76
[ba] = + 5 002.60	[bb] = + 2 777.05	[bc] = + 17 709.99	[bs] = + 25 489.64

[a] We again have: $b_i a_i + b_i b_i + b_i c_i = b_i s_i$ and $[ba] + [bb] + [bc] = [bs]$.

The normal equations will be, with the coefficients rounded off,

$$11\,997x + 5\,003y = +29\,855$$
$$5\,003x + 2\,777y = +17\,710$$

The solutions are

$$x = -0.6874$$
$$y = +7.6158$$

By formulae (III.5), p. 35 it is found

$$A = 95°.2 = 6^h21^m$$
$$D = +7°.5$$

From the Equations (III.5) one easily finds

$$\sin A = y/(x^2 + y^2)^{1/2}, \quad \cos A = x/(x^2 + y^2)^{1/2},$$

which gives

$$\sin A = +0.9960, \quad \cos A = -0.0899.$$

We also have

$$\sin D = (1 + x^2 + y^2)^{-1/2}, \quad \cos D = [(x^2 + y^2)/(1 + x^2 + y^2)]^{1/2},$$

which gives

$$\sin D = +0.1297, \quad \cos D = +0.9916.$$

Make use of the formula

$$\cos(A - \alpha) = \cos A \cos\alpha + \sin A \sin\alpha$$

to compute $\cos(A - \alpha)$ for all the stars. Thereafter find $\cos\lambda$ by formula (III.18) and, finally, $\cos\theta_{calc}$ by formula (III.19).

TABLE III.6

vB	$\cos(A - \alpha)$	$\cos\lambda$	λ	$\cos\theta_{calc}$	θ_{calc}	θ_{obs}
6	+ 0.7925	+ 0.7891	37°.9	− 0.1762	100°.1	100°.6
14	+ 0.8388	+ 0.8403	32 .8	+ 0.0940	84 .6	86 .2
20	+ 0.8487	+ 0.8460	32 .2	− 0.1813	100 .5	103 .5
41	+ 0.8648	+ 0.8570	31 .0	− 0.2583	105 .0	105 .5
60	+ 0.8719	+ 0.8476	32 .0	− 0.4033	113 .8	113 .1
74	+ 0.8776	+ 0.8772	28 .7	− 0.1425	98 .2	96 .1
107	+ 0.8984	+ 0.9002	25 .8	+ 0.0183	88 .9	88 .8
108	+ 0.8984	+ 0.8924	26 .8	− 0.2616	105 .2	101 .2
123	+ 0.9202	+ 0.9057	25 .1	− 0.4025	113 .7	112 .3
130	+ 0.9483	+ 0.9487	18 .4	− 0.1001	95 .7	92 .6

For the (absolute value of the) space velocity one easily finds $V = 41.8$ km/s.

SOLAR MOTION AND VELOCITY DISTRIBUTION OF
A GROUP OF STARS

1. From the standpoint of stellar kinematics the type of motion shown by the members of the moving clusters – of which the Hyades are the best-known example – represents rather an exception than the rule. In fact, if we take an arbitrary point in space and consider all the stars contained in an astronomically small volume around this point, we shall find that they move in all directions and that the absolute values of their velocities scatter over a wide range. The motion of the members of a moving cluster can be considered as perfectly ordered. At first sight it seems that just the opposite type of perfectly random motion is exhibited by the large majority of common stars.

The motion of a group of stars forming a moving cluster can fully be described by the space velocity of the cluster, this quantity – considered always as a vector – having the same value for all cluster members. In the present exercise we shall investigate how it is possible to characterise quantitatively the motion of stars contained in an arbitrary chosen elementary volume of space in our stellar system. Obviously, in a first approach we shall not attempt to arrive at a description embracing all details, but rather to disclose and then to describe or, even better, to characterise by suitably chosen quantities, some of the most salient features in the motions of the stars. We may, in fact, say that our problem belongs to the domain of statistics.

Even if we did not know anything about stellar motions, we could expect that there must exist a relation between the geometric structure of our Galaxy and the motions of its members. In fact, the distribution of different classes of objects in our Galaxy, and the structure of the Galaxy as a whole, as recorded by observations, represent only a snapshot. If, now, we suppose that in the course of time our stellar system does not change too rapidly, then we can well assume that the present distribution remains preserved for a certain period of time. This notwithstanding the fact that, due to their motions, other individuals (belonging to the same classes) have occupied, or will approximately take the positions held by the objects we observe at this time. This however means that, e.g. the velocity components perpendicular to the galactic plane must be smaller for objects exhibiting a strong concentration towards the galactic plane than for those which can be found at large distances from it. In other words we can in general expect that objects belonging to different classes as judged by their spatial distribution, will also have different motions. Due to the well-known link between the spatial distribution and the physical properties we can even expect that physically different objects will exhibit different kinematical properties. It is therefore clear that, if our investigations of the motions of stars are to lead to meaningful results, we certainly shall not at once consider all stars contained in a given volume element, but instead investigate separately the different classes taking always only such individuals which form a homogeneous group.

The origin of the questions we will discuss can be traced back to the time when the first determinations of the solar apex were made. In those early days it was the motion of the Sun itself which attracted the attention of the astronomers. Later the interest shifted to the relative motions of different classes of objects, the Sun being degraded to the role of an intermediary reference point. The investigations of the very nature of the randomness of stellar motions came even later, opening new problems some of which have not yet found a solution.

2. When investigating, later in this Exercise, the motion of a group of stars, we shall consider their space velocities as given quantities. Now the space velocity of a star cannot be determined directly. Instead it is deduced from other, observable, quantities. In order to get a better insight into the real meaning of what we call a space velocity, and also in order to be able to appreciate (at least some of) the effects which may systematically influence the values at our disposal, let us see how the space velocities are defined and determined.

The answer to this question is given by the Equations (III.7), (III.8) and (III.9), to which the reader is referred. In particular we have:

$$
\begin{aligned}
- X \sin\alpha \quad\quad + Y \cos\alpha \quad\quad\quad\quad\quad\quad &= V_\alpha = 4.738 r \mu_\alpha \cos\delta_i \\
- X \sin\delta \cos\alpha_i - Y \sin\delta \sin\alpha + Z \cos\delta &= V_\delta = 4.738 r \mu_\delta \quad\quad\quad\text{(IV.1)} \\
+ X \cos\delta \cos\alpha + Y \cos\delta \sin\alpha + Z \sin\delta &= \varrho .
\end{aligned}
$$

The quantities V_α, V_δ, and ϱ, are the components of the space velocity along the axes of the astronomical trihedral. The quantities X, Y, and Z, are the components of the space velocity, referred to the system of equatorial space coordinates (that is to a system having its origin in the Sun), with the x-axis pointing towards the vernal point, the y-axis towards the point $\alpha = 6$ h, $\delta = 0°$, and the z-axis towards the North Celestial Pole, for a given epoch). If for the star we know the equatorial coordinates (α, δ), the proper motion components (μ_α, μ_δ), the radial velocity (ϱ) and the distance (r), then we can solve Equation (IV.1) for X, Y, and Z.

The solution, which can immediately be written down, is:

$$
\begin{aligned}
X &= - 4.738 r \mu_\alpha \cos\delta \sin\alpha - 4.738 r \mu_\delta \sin\delta \cos\alpha + \varrho \cos\delta \cos\alpha \\
Y &= + 4.738 r \mu_\alpha \cos\delta \cos\alpha - 4.738 r \mu_\delta \sin\delta \sin\alpha + \varrho \cos\delta \sin\alpha \quad\text{(IV.2)} \\
Z &= \quad\quad\quad\quad\quad\quad\quad\quad + 4.738 r \mu_\delta \cos\delta \quad\quad + \varrho \sin\delta .
\end{aligned}
$$

The proof is left to the reader (see Problem IV.1 at the end of this exercise).

When the distances are measured in parsecs, the components of the proper motion expressed in seconds of arc per year, and the radial velocity in km/s, Equations (IV.2) give the components of the star's velocity in km/s.

It is important to notice that the space velocity, as defined by these components, is in fact the relative velocity referred to the Sun. In this sense it can be called heliocentric.

Notice, further, that the components X, Y, and Z, refer to an equatorial frame whose axes are related to the Earth's rotation axis and its orbital plane. The choice of this frame is imposed by the methods of observation. For our present purposes, however, such a frame cannot be considered as an especially suitable one. In fact, it is obvious that regularities eventually present in the motions of the stars, will more easily be recognised if components related to the axes of the new system of galactic coordinates are used. By analogy we could take, for a system of galactic space coordinates, a first axis pointing towards the galactic center ($l=0°$, $b=0°$), a second directed towards the point $l=90°$, $b=0°$, with the third pointing towards the North Galactic Pole. In practice, however, a somewhat different system is often adopted, the sole difference being that the first axis is directed towards the anti-center ($l=180°$, $b=0°$). The components along these axes are denoted by U, V, and W, or by Π, Θ, and Z, respectively. In the present exercise we will adopt the former notation (see Figure IV.1).

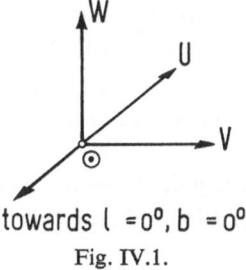

towards $l=0°$, $b=0°$

Fig. IV.1.

The components U, V, and W, can be computed from the (equatorial) components X, Y, and Z, by the following transformation formulae:

$$\begin{aligned}
U &= + 0.06699X + 0.87275Y + 0.48354Z \\
V &= + 0.49273X - 0.45034Y + 0.74459Z \\
W &= - 0.86760X - 0.18838Y + 0.46020Z.
\end{aligned} \qquad \text{(IV.3)}$$

The proof is left to the reader (Problem IV.2).

From the Equations (IV.2) it is seen that we need a very complete set of data in order to be able to compute the space velocity of a star (α, δ; μ_α, μ_δ; ϱ; r). Now the number of stars for which all these data (and especially the proper motions and the distance) are known with sufficient accuracy is relatively small. This means that we will be able to investigate only a small portion, a *sample* as it is called, of all the stars belonging to a given class. If now from the results of our investigation we will draw some conclusions concerning the behavior of the whole class – and this is certainly our intention – then we must ask whether or not our sample can be representative for the whole. If, for the distances one relies on trigonometric parallaxes, one risks favoring stars with large proper motions, because such stars are more often put on observational programs for parallax determination. In this case our sample will be *biased by an effect of selection*. But even if this is not the case, sufficiently accurate proper motions can be derived only for relatively nearby stars. This means that, in fact, we cannot arbitrarily choose a volume element in space but that, instead, our investigation will be restricted to stars in the vicinity of the Sun. Parallaxes are probably the most important source of uncertainty in the determination of space velocities. One can hope that, with the advent of more accurate methods of photometric distance determinations, the situation will improve. But even then the restriction imposed by the proper motions will not be removed. In brief, the study of the motions of the stars by means of the space velocities certainly has many drawbacks. On the other hand it has the great advantage of being the simplest from a theoretical standpoint. Therefore it is this method that we will adopt in the present exercise.

In order to circumvent the difficulties connected with the determination of space velocities, other methods, based only on radial velocities, or only on proper motions, have been devised. For an account of these methods as well as for the discussion of other points raised in the preceding paragraph the reader is referred to the monographs given under [1], [2] and [3].

The most recent and authoritative account of the investigations concerning the motions of the common stars is that given by J. Delhaye in Vol. V., *Galactic Structure*, of the *Compendium of Astronomy and Astrophysics* [4].

The most recent and complete general catalogue of space velocities is that compiled by O. J. Eggen [5].

Very useful tables permitting the galactic velocity components to be found from the observed ones have been prepared by L. Perek [6].

3. Table IV.1 (p. 64) contains the components U, V, and W of 75 nearby early F-type stars belonging to the main sequence. What shall we do to arrive at a simple quantitative description of the most salient features of the motions of these stars? Usually it is admitted that this has been achieved if the following questions can be be answered:

(1) What is the mean velocity of the stars, and,

(2) How are the individual velocities distributed around this mean.

In order to answer the first question we have to compute the vector of the mean velocity of the stars concerned. For the second one we will have to produce and interpret suitable graphs.

The vector of the mean velocity is defined by its components $\bar{U}$, $\bar{V}$, and $\bar{W}$, which are given by

$$\bar{U} = \frac{1}{n} \sum_{i=1}^{n} U_i, \qquad \bar{V} = \frac{1}{n} \sum_{i=1}^{n} V_i, \qquad \bar{W} = \frac{1}{n} \sum_{i=1}^{n} W_i, \qquad (IV.4)$$

where U_i, V_i, and W_i, are the components of the space velocity of the i-th star, n being the total number of stars. The point moving at the mean velocity of the group is called the centroid of the group.

The coordinates of the solar apex can be found from $\bar{U}$, $\bar{V}$, and $\bar{W}$, as follows. By definition, the solar apex (with respect to a given group of stars) is the point towards which the Sun is moving. Now if $\bar{U}$, $\bar{V}$, and $\bar{W}$, are the components of the mean velocity, with respect to the Sun, of the stars considered, the $-\bar{U}$, $-\bar{V}$, and $-\bar{W}$, are the components of the Sun's velocity with respect to this same group and referred to the same axes as $\bar{U}$, $\bar{V}$, and $\bar{W}$. In order to compute the galactic coordinates of the apex, that is of the point towards which $-\bar{U}$, $-\bar{V}$, and $-\bar{W}$, is directed, remember that the U components are referred to an axis pointing towards $l = 180°$, $b = 0°$, whereas the first axis of the galactic space coordinate system is directed towards the galactic center ($l = 0°$, $b = 0°$). Therefore, denoting the galactic longitude and the galactic latitude of the solar apex respectively by l_A and b_A, we will have:

$$\text{tg}\, l_A = -\bar{V}/\bar{U}, \qquad \sin b_A = -\bar{W}/S \qquad\qquad (\text{IV.5})$$

where
$$S = [\bar{U}^2 + \bar{V}^2 + \bar{W}^2]^{1/2}\ \text{km/s} \qquad\qquad (\text{IV.6})$$

is the absolute value of the Sun's velocity relative to the stars considered. Usually S is called the Sun's velocity, although it is not a vector.

The right ascension and the declination of the apex can be found from l_A and b_A by transformation formulae or, with sufficient accuracy, from transformation tables or even nomograms.

The quantities l_A, b_A, and S, are sometimes called the elements of the solar motion (with respect to a group of stars).

Using the velocities given in Table IV.1 compute the mean velocity of the group with respect to the Sun, as well as the elements of the solar motion. Locate on a celestial map the position of the apex.

Sufficiently accurate values of $\bar{U}$, $\bar{V}$, and $\bar{W}$, can very easily be found by a method which we shall explain taking the determination of $\bar{U}$ as an example.

Find, by inspection of Table IV.1, the maximal value U_M and the minimal value U_m of the component U. Now take a suitable round value U_{mo} somewhat lower than U_m (e.g. -40.0 if $U_m = -37.2$), and a round value U_{Mo} somewhat larger than U_M (e.g. $+50.0$, if $U_M = +46.7$). Then divide the interval $U_{Mo} - U_{mo}$ into a number of sub-intervals ΔU of equal length (taking, for example, $\Delta U = 10.0$) beginning with the value 0.0, and proceeding towards larger as well as towards smaller values of U. Let now U_k ($k = \pm 1$, ± 2, ± 3, ...) be the values corresponding to the center of each interval (which, in our example will all be uneven multiples of ± 5). Let, further, n_k be the number of stars with U-components falling within the sub-interval centered on U_k (or, what amounts to the same: the number of U-components present in this interval). Then, as a good approximation for $\bar{U}$ we can take

$$\left(\sum_k n_k U_k \right) \bigg/ \sum_k n_k. \qquad\qquad (\text{IV.7})$$

The other two components of the mean velocity vector are given by analogous formulae which can be derived in the same way.

4. Having derived the mean velocity we now must see how the distribution of the individual velocities around this mean can be described. As we shall see, the histograms of U, V, and W components will not serve the purpose.

A simple graphic method which is currently used, and which can be refined, is the following one. Choosing suitable units consider the components U_i, V_i, W_i, as the coordinates, with respect to the axes U, V, W, of a representative point. By this means the velocities of the group of stars will be represented by a cloud of points. The distribution of the individual velocities can then be derived from, and described in terms of, the geometry of the cloud. Notice that the point U_i, V_i, W_i, is the end point of a vector parallel to, and of the same intensity as, the space velocity of the i-th star, but having its origin in the Sun.

Now it is somewhat difficult to make a three-dimensional representation of such a cloud of points. However its properties can easily be investigated from the three projections onto the three principal planes, which are:

(1) The plane containing the axes U and V, or the symmetry plane of the Galaxy;

(2) The plane perpendicular to it, containing the axes U and W. This plane is identical to the plane through the symmetry axis of the Galaxy and the position of the Sun, and is called the meridional plane;

(3) The plane perpendicular to both of them, drawn through the axes V and W.

For this purpose make three separate diagrams, first, one UV-diagram where all the points U_i, V_i, are plotted, then a UW-diagram containing the points U_i, W_i, and finally the VW-diagram (with the points V_i, W_i).

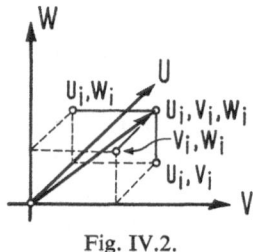

Fig. IV.2.

Obviously, the point U_i, V_i represents the projection, onto the UV plane, of the end point of the space velocity vector U_i, V_i, W_i, or, what amounts to the same, the end point of that velocity component of the i-th stars which is in the UV plane. The set of points U_i, V_i ($i=1, 2, ..., n$) will therefore give a geometric representation of the distribution of the velocities in the symmetry plane. Similarly the UW and the VW diagrams will represent the velocity distribution in the UW and VW planes, respectively.

Notice that on all the diagrams the zero point corresponds to the velocity of the Sun, which is due to the fact that the space velocities used are relative velocities, referred to the Sun. Suppose now that we wish to investigate the velocity distribution with respect to another reference point, moving relative to the Sun at a velocity U_p, V_p, W_p. The Sun's velocity relative to this point will be $-U_p$, $-V_p$, $-W_p$; that of the i-th star, relative to the Sun, is U_i, V_i, W_i. Therefore, the i-th star moves relatively to the new reference point at the velocity U_i-U_p, V_i-V_p, W_i-W_p. The velocity distribution will now be represented by points having the coordinates $U_i - U_p$, $V_i - V_p$, $W_i - W_p$. Therefore the new diagrams can be obtained from the old ones simply by shifting the zero points from their initial position into the points which, in the respective diagram, represents the velocity of the new reference point.

Velocities referred to the centroid are called residual velocities. By definition the components of the residual velocity of the i-th star will be

$$U_i - \bar{U}, \quad V_i - \bar{V}, \quad W_i - \bar{W}.$$

In order to get the diagrams representing the distribution of the residual velocities we have, therefore, to shift the zero point of each of the diagrams from its original position into $\bar{U}$, $\bar{V}$, resp. $\bar{U}$, $\bar{W}$, resp. $\bar{V}$, $\bar{W}$.

Make for all the stars listed in Table IV.1 the three diagrams. Figure IV.3 represents schematically the kind of diagrams you should get.

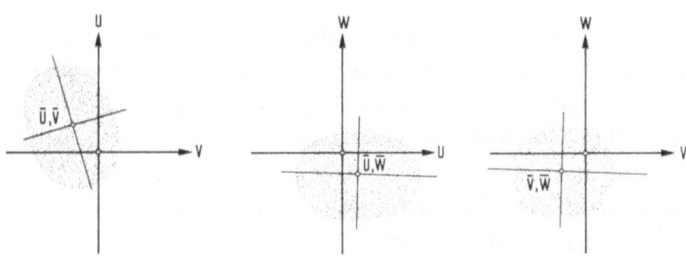

Fig. IV.3.

5. What are the mean features of the velocity distributions in the three principal planes?

On both the UV and the UW diagrams the points cover an area which is very markedly elongated; in the VW plane the elongation is not so pronounced. We infer therefrom that the cloud of representative points itself must have an elongated form. On both the UV and the UW diagrams one could trace, through the centers of these distributions (that is, through the points $\bar{U}$, $\bar{V}$, and $\bar{U}$, $\bar{W}$ respectively) the respective major axes. In principle this could also be done on the VW diagram. However, as this distribution is not so markedly elongated, the orientation of its major axis will not be so sharply defined as for the previous two.

The major axis of the UV distribution deviates markedly from the direction of the U-axis. On the other hand the major axis of the UW distribution coincides practically with the U-axis. Now the UV and the UW distributions are the projections, on the UV and the UW planes, respectively, of the representative points U_i, V_i, W_i. Therefore we must conclude that the cloud of representative points itself has a major axis which is practically parallel to the UV plane, but which deviates from the direction of the U-axis, or, what amounts to the same, from the direction towards the galactic center.

The distribution in the VW plane does not contradict such a conclusion. It shows, moreover, the existence of a minor axis which is very nearly parallel to the W axis, corresponding to the direction normal to the galactic plane. Finally, one can identify a third, intermediary axis, seen in projection on the UV plane as the minor axis of the UV distribution. As far as one can guess from the VW diagram, this axis is nearly parallel to the UV plane.

How can such a distribution of the representative points be interpreted?

According to the geometric method of representation which we have adopted, the coordinates of any point are equal to the components of the space velocity of a given star along the coordinate axes. Now we are free to choose not only the zero point but also the direction of the axes. As the new zero point we will choose the point corresponding to the velocity of the centroid, i.e. the point $\bar{U}$, $\bar{V}$, $\bar{W}$, or, on our three diagrams, the points $\bar{U}$, $\bar{V}$, and $\bar{U}$, $\bar{W}$, and $\bar{V}$, $\bar{W}$ respectively. Let us further take, as a new U-axis, which we shall denote by U', the major axis of the cloud of representative points. The fact that in this direction the points are more spread out than in any other means that the components of the residual velocities along the U'-axis are, by absolute value, and in the mean, larger than in any other direction. We can also say that in the direction defined by this axis the motion of the stars show a larger scatter than in any other. In this sense this axis defines a privileged direction in the distribution of the residual velocities. The two diametrically opposite points on the celestial sphere corresponding to this direction are called the vertexes of the group of stars considered.

What can we do to characterise quantitatively the mean features of such a distribution of the residual velocities? We shall

First, specify the directions of the axes of the distribution, and,

Second, characterise, by a suitably chosen quantity, the magnitude of the scatter of the components corresponding to each of these axes.

This can be achieved either by graphic or by numerical methods. For the sake of clarity and simplicity let us adopt a simple graphic method giving reasonably good approximate results. A more refined numerical method will be shown in the following exercise.

Concerning the first point let us focus our attention on the major axis of the distribution. As already said, this axis is very approximately in the UV plane. Neglecting a small inclination, we shall specify its direction by the galactic longitude of one of the two vertexes. Usually one takes the point which on the celestial sphere is closest to the galactic center.

On the UV diagram trace, through the center of the distribution, i.e. through the point $\bar{U}$, $\bar{V}$, the major axis as well as the minor axis perpendicular to the first. Check the orientation of the axes by counting the points in the four quadrants. Assume that the axes are properly oriented if each quadrant contains very nearly the same number of points. Improve the orientation of the axes if necessary. Determine the angle between the major axis and the original U-axis. It is this angle which is usually called the vertex deviation. We shall denote it by l_V.

Concerning the second point let us adopt, as the measure of the scatter around the central value, the dispersion of the relevant component.

Denote by U_i' the component of the residual velocity of the i-th star along the U'-axis. By definition the dispersion of the U' components is given by

$$\sigma_{u'} = \left[\frac{1}{n} \sum_{i=1}^{n} (U_i')^2\right]^{1/2}.$$ (IV.8)

The quantity under the square root is the mean of the squared U' components. For this mean an acceptable approximate value can be found as follows, by a method quite analogous to that used in Section 3 to derive $\bar{U}$, $\bar{V}$, and $\bar{W}$.

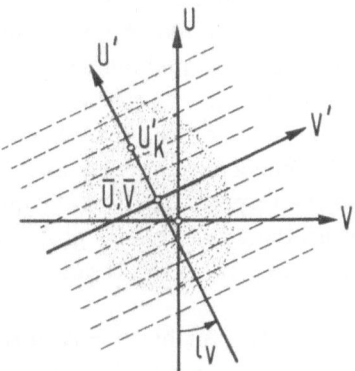

Fig. IV.4.

On the UV diagram divide the U'-axis into segments of equal length, say $\Delta U' = 10$ km/s beginning from the zero point of the U'-axis and proceeding in both the positive and the negative directions (Figure IV.4). By this means the whole UV plane will be divided into strips of equal width. Let U_k' ($k = \pm 1, \pm 2, ...$) be the values corresponding to the mid points of each segment, which in our case will correspond to positive and negative uneven multiples of 5. Let, further, n_k be the number of points contained in the strip centered on U_k. Then, as a good approximation of the mean of the squared U_i' ($i = 1, 2, 3, ..., n$) we can take

$$\left(\sum_k n_k U_k'^2\right)\Big/\sum_k n_k.$$ (IV.9)

Determine in this way the dispersion of the component along the major axis as well as the dispersion σ_V, of the components along the minor axis V' perpendicular to the first. Obviously the same method can be used to find from the VW diagram the dispersion σ_W, of the components along the minor axis of the cloud of representative points, assuming that this axis is parallel to the W axis.

Compare the three dispersions and compute their ratios.

Using the values n_k make the corresponding histogram for each component of the residual velocity for which the dispersion has been determined.

Summarise the results by stating:

The number of stars used as well as their spectral type;

The galactic coordinates of the apex (l_A, b_A) and the Sun's velocity S km/s;

The vertex deviation (l_V) and the dispersions.

Compare your values to those which can be found in Table 1 in Delhaye's article. Let us add some closing remarks.

In this exercise we have been interested in the motion of the bulk of common stars in the solar neighbourhood. We have considered the type of motion exhibited by the moving clusters as an exception. Accordingly we have excluded from our list, Table IV.1, all members of the known moving clusters.

Now it is possible that the real situation does not permit such a clearcut distinction nor such a simple treatment. In fact, making use of more accurate data than previously available, O. J. Eggen has shown that apart from the known members of some moving clusters, there are many other stars which share their motion although on the sky they lie far from the clusters concerned. Such stars form what Eggen calls the groups corresponding to the respective clusters. To give an example, apart from the well known Hyades cluster, there is a Hyades group, consisting of stars scattered over the sky, which share the motion on the Hyades so closely that this cannot be considered as pure chance.

However Eggen has made a decisive further step by formulating the following question: "If the majority of the stars were produced in a few batches, and if the disruptive forces (of the galactic gravitational field) are not greatly effective, might not the individual stars of a batch still be identifiable by their motion?". As to the consequences, let once more quote Eggen: "If so, the space motions near the Sun should be distributed in a non-random way. It is usual in applying the various statistical procedures used in the study of stellar motions to assume that these motions are randomly distributed with, at most, minor variation. If in fact the observed motions are dominated by those of a relatively few stellar groups, then many of these procedures may be invalid." But Eggen's surmise concerning the character of the stellar motion has even wider implications: "Furthermore, if the now widely spread members of an original bath of stars can be identified by their motion, a large sample of coeval stars could be examined for chemical constitution and distribution on the color-luminosity array and the possibility of catching stars in such interesting, rapid stages of their evolution as the Hertzsprung gap would be greatly increased".

In a long series of publications Eggen has given evidence of the existence of many such groups of stars (not associated with moving clusters). It is, however, beyond the scope of the present exercise to enter this question. The interested reader is referred, for a general review of this problem, as well as for references on previous work, to Eggen's article in Vol. V of the series *Stars and Stellar Systems*, p. 7. More recent results and references can be found in two of Eggen's articles published in the *Publications of the Astronomical Society of the Pacific* and in the *Astrophysical Journal* (**8, 9**).

Problem IV.1. Derivation of the Formulae for the Equatorial Components of the Space Velocity

Solve the Equations (IV.1) for X, Y, and Z.

Solution

The Equations (IV.1) represent transformation formulae permitting passage from the components of the space velocity referred to the equatorial space coordinate system to the components of the same vector referred to the axes of the astronomical trihedral. Disregarding a translation of the origin which has no physical meaning, any of the two coordinate systems can be derived from the other one by rotation. In other words Equations (IV.1) are transformation formulae for rotation of axes. In this case, as shown in any textbook of vector calculus, the matrix of the inverse transformation, giving the components X, Y, and Z, is simply obtained by transposition, i.e. by interchanging the rows and the columns of the initial matrix. This immediately gives the solutions (IV.2).

Problem IV.2. Derivation of the Formulae for the Galactic Components of the Space Velocity

Derive the transformation formulae (IV.3).

Solution

In order to prove Equations (IV.3) notice first that the system of axes, to which the components U, V, and W, refer, differs from the new system of galactic space coordinates only by the orientation of its first axis. Notice, further, that the origin of the new galactic coordinate system coincides with the origin of the space equatorial coordinate system, so that their respective axes can be brought to coincidence by suitable rotations.

The derivations of Equations (IV.3) will therefore consist of two steps. First we have to bring the equatorial axes into coincidence with the axes of the new galactic coordinate system. This being achieved we have only to change the direction of the first axis.

The first step will consist of three successive rotations. The angles of rotation are determined by the directions of the axes of the new galactic coordinate system with respect to the axes of the equatorial frame (for a given epoch). They are fixed by the definition of the new galactic coordinate system which states that, with respect to the equatorial frame for the epoch 1950.0, the new polar galactic axis is directed towards the point $\alpha_p = 12$ h 49 m, $\delta_p = +27.4°$, whereas the first (or 'x') axis has a position angle at this pole which is equal to 123°. The second (or 'y') galactic axis is normal to both of them and oriented in such a way that all three form a right-handed system.

First rotate the initial equatorial frame around its polar axis by an angle of $12\,\mathrm{h}\,49\,\mathrm{m} = 192°\,15'$ in the positive direction, see Figure IV.5a which corresponds to the plane of the celestial equator. According to transformation formulae well-known from analytic geometry, and of the same type as those already used in Problem I.2 (Equations (P I.16)), the matrix corresponding to this rotation is

$$M_1 = \begin{pmatrix} +\cos 192°15' & +\sin 192°15' & 0 \\ -\sin 192°15' & +\cos 192°15' & 0 \\ 0 & 0 & 1 \end{pmatrix} = \begin{pmatrix} -0.97723 & -0.21218 & 0 \\ +0.21218 & -0.97723 & 0 \\ 0 & 0 & 1 \end{pmatrix}. \quad (\text{P.IV.1})$$

Rotate now this new frame around its second ('y') axis in the negative direction by $90° - 27.4° = 62.6°$. This second rotation will bring its polar ('z') axis into coincidence with the new galactic polar axis see Figure IV.5b which represents the plane drawn through both the North Celestial and the North Galactic Poles and the origin. The matrix corresponding to the second rotation will be

$$M_2 = \begin{pmatrix} +\cos 62.4° & 0 & -\sin 62.4° \\ 0 & 1 & 0 \\ +\sin 62.4° & 0 & +\cos 62.4° \end{pmatrix} = \begin{pmatrix} +0.46020 & 0 & -0.88782 \\ 0 & 1 & 0 \\ +0.88782 & 0 & +0.46020 \end{pmatrix}. \quad (\text{P IV.2})$$

Finally rotate the frame obtained around its polar axis by $180° - 123° = 57°$ in the

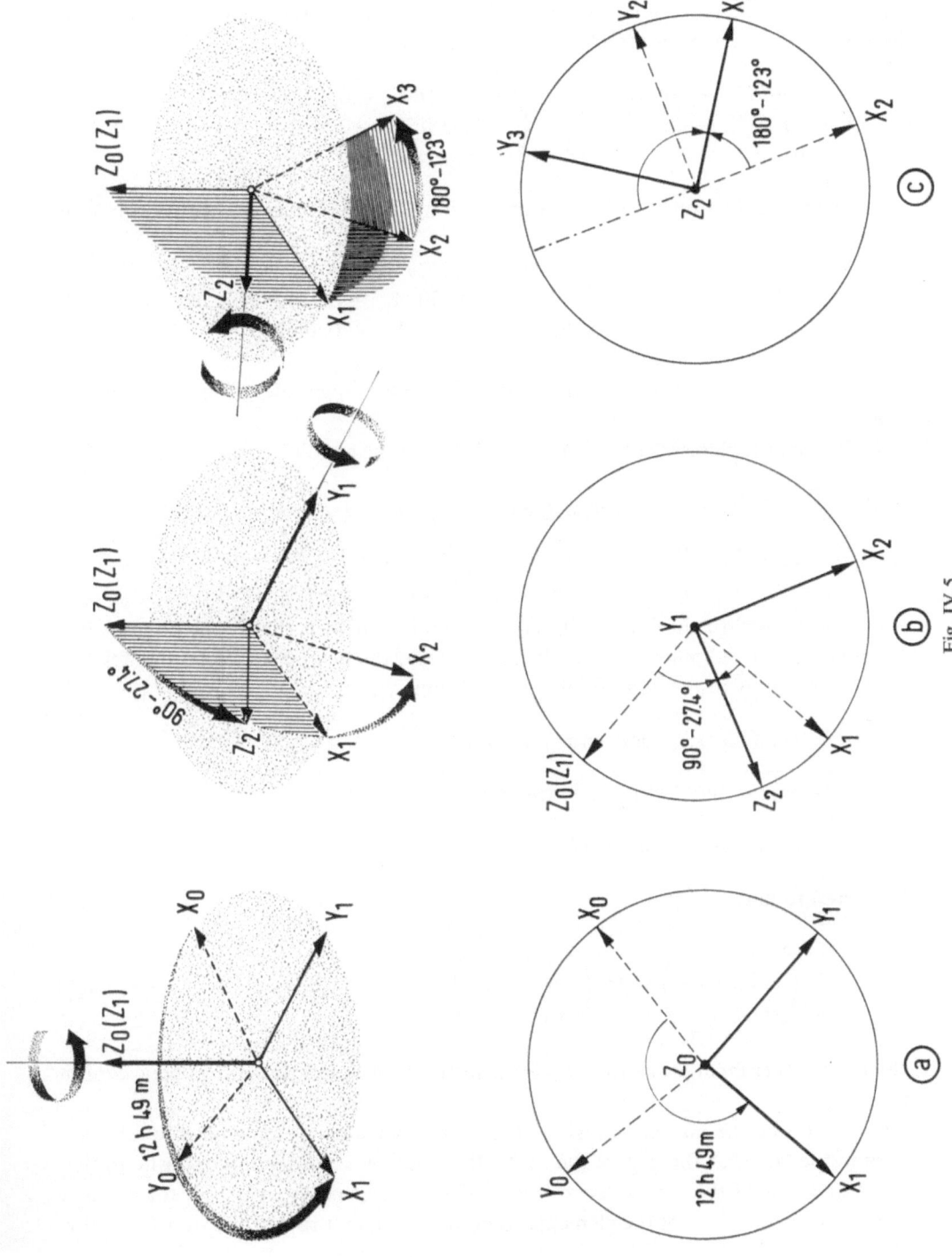

Fig. IV.5.

positive direction. This third rotation will bring all its axes into coincidence with the new galactic axes, see Figure IV.5c which corresponds to the plane of the galactic equator. The matrix corresponding to this rotation is

$$M_3 = \begin{pmatrix} +\cos 57° & +\sin 57° & 0 \\ -\sin 57° & +\cos 57° & 0 \\ 0 & 0 & 1 \end{pmatrix} = \begin{pmatrix} +0.544\,64 & +0.838\,67 & 0 \\ -0.838\,67 & +0.544\,64 & 0 \\ 0 & 0 & 1 \end{pmatrix}. \qquad \text{(P.IV.3)}$$

The matrix of the transformation corresponding to these three consecutive rotations is

$$M_1 \cdot M_2 \cdot M_3 = \begin{pmatrix} -0.066\,99 & -0.872\,75 & -0.483\,54 \\ +0.492\,73 & -0.450\,34 & +0.744\,59 \\ -0.867\,60 & -0.188\,38 & +0.460\,20 \end{pmatrix} = M, \qquad \text{(P.IV.4)}$$

a result found after some computations, using the well-known rules for matrix multiplication.

Notice that the elements of this matrix must fulfil the conditions of orthonormality:

$$\sum_k a_{ik}^2 = 1, \quad i = 1, 2, 3; \qquad \sum_i a_{ik}^2 = 1, \quad k = 1, 2, 3;$$

$$\sum_k a_{ik} a_{jk} = 0, \quad i, j = 1, 2, 3; \qquad \sum_i a_{ik} a_{ij} = 0, \quad k, j = 1, 2, 3.$$

which can be used to check the values found. Notice, further, that to the point having the equatorial coordinates (epoch 1950.0) α, δ, and the new galactic coordinates l, b, there corresponds in the equatorial frame the unit vector

$$\cos\delta \cos\alpha, \quad \cos\delta \sin\alpha, \quad \sin\delta.$$

Its components in the new galactic frame are

$$\cos b \cos l, \quad \cos b \sin l, \quad \sin b.$$

We therefore have

$$\begin{pmatrix} \cos b \cos l \\ \cos b \sin l \\ \sin b \end{pmatrix} = M \cdot \begin{pmatrix} \cos\delta \cos\alpha \\ \cos\delta \sin\alpha \\ \sin\delta \end{pmatrix}, \qquad \text{(P.IV.5)}$$

which is another form of the formula for transforming equatorial into galactic coordinates.

Having found the matrix corresponding to the rotations, we only have to take into account the fact that the first of the U, V, W system has a direction opposite to that of the first axis of the new galactic frame. This is most easily achieved, for we have only to invert the signs of the elements standing in the first row of the transformation matrix.

References

[1] Smart, W. M.: 1968, *Stellar Kinematics*, Longmans, Green and Co., London.

[2] Ogorodnikov, K. F.: 1965, *Dynamics of Stellar Systems*, Pergamon Press, Oxford.

[3] Trumpler, R. J. and Weaver, H. F.: 1953, *Statistical Astronomy*, University of California Press, Berkeley. Reprint: Dover Publications, New York, 1962.

[4] Delhaye, J.: 1965, in A. Blaauw and M. Schmidt (eds.), *Galactic Structure*, University of Chicago Press, Chicago.

[5] Eggen, O. J.: 1962, 'Space-Velocity Vectors for 3483 Stars', *Roy. Obs. Bull.* No. 51, HM Stationery Office, London.

[6] Perek, L.: 1964, 'Tables of Galactic Direction Cosines', Czechoslovak Academy of Sciences, Astronomical Institute Publication No. 48, Prague.

[7] Eggen, O. J.: 1965, in A. Blaauw and M. Schmidt (eds.), *Galactic Structure*, University of Chicago Press, Chicago.

[8] Eggen, O. J.: 1969, *Publ. Astron. Soc. Pacific* **81**, 553.

[9] Eggen, O. J.: 1969, *Astrophys. J.* **155**, 701.

TABLE IV.1[a]

Space velocities of 75 early F-type stars

BS	U	V	W	BS	U	V	W
21	+31.4	− 4.2	−16.8	4421	+16.2	+18.9	−29.0
41	−33.3	−21.2	− 6.7	4480	+ 0.6	−20.4	−22.6
230–1	+27.3	−13.3	− 5.6				
327	+37.9	−14.5	−10.7	4825–6	+28.5	− 6.0	−18.2
330	−40.6	+ 6.5	−22.5	5050	−23.3	+ 4.1	+ 6.7
				5083	−24.9	−14.3	+ 1.9
368	+20.6	+29.0	− 7.6	5245	−12.4	+24.2	−30.3
413	− 6.3	− 1.8	+ 6.4	5365	− 9.7	+ 5.3	− 6.6
529	− 5.4	−20.9	−11.8				
638	−15.4	−19.7	−20.8	5434	+28.4	−14.6	+ 0.3
673	−26.8	+ 3.7	+ 9.5	5447	− 2.2	+16.5	− 5.4
				5529	+43.5	−29.1	−15.9
765	+28.8	−12.4	+ 0.7	5537	− 1.5	− 0.5	− 5.8
770	+16.6	−14.8	− 6.9	6093	+45.5	−16.8	− 9.2
783	+17.6	− 3.3	− 9.9				
878	+38.2	−16.8	− 2.1	6181	+ 8.2	−21.6	−10.9
988	+ 1.2	+ 7.3	+ 6.4	6493	− 1.9	−11.4	+ 7.4
				6594	+41.2	−16.6	−12.3
1210	+35.4	− 8.5	− 6.8	6670	+31.9	−17.7	+ 8.5
1218	+31.4	−18.6	−17.5	6710	+37.7	−12.8	−22.5
1269	+ 1.2	+ 2.6	−13.2				
1276	+ 6.3	−19.9	− 2.6	6849	+12.9	− 9.4	+ 4.2
1287	+14.9	+ 1.9	−15.5	6850	+ 4.3	−17.6	−42.2
				6985	+ 7.1	−24.1	− 9.2
1637	+ 5.9	−13.3	−14.4	7044	−25.5	−15.3	− 6.3
2085	+ 4.7	+ 7.4	+ 1.8	7261	+12.7	+15.6	−22.5
2150	+27.4	−17.5	− 4.4				
2241	+12.3	+ 7.9	+14.3	7266	+28.4	−38.9	− 3.3
2264	+ 8.4	−23.3	− 3.3	7460	+33.5	−22.4	− 8.1
				7469	+20.1	−27.3	+ 3.1
2530	−24.9	−11.0	− 8.7	7495	+37.0	−11.8	− 8.5
2777	+ 2.7	− 1.2	− 1.3	7697	+50.4	−36.4	−14.8
2852	− 9.6	+ 9.4	+11.6				
2930	+10.0	−20.1	− 8.2	8330	+18.6	−20.1	− 6.4
3087	+13.5	−21.9	0.0	8376	− 2.5	−17.3	−14.4
				8392	+ 2.1	− 5.5	−22.2
3106	−33.6	− 9.3	−19.0	8666	+31.4	−18.2	− 7.5
3254	−16.9	− 3.7	− 1.3	8718	+ 9.1	−29.4	+ 4.3
3624	−13.1	−11.2	+ 6.9				
3857	− 3.9	− 9.0	− 8.2	8735	+24.1	−17.8	− 1.3
4084	+ 9.5	+ 4.9	− 2.1	8977	+38.5	−15.8	− 7.4
				9020	+17.8	+24.5	− 5.2
4230	+ 5.3	+ 6.3	− 1.7	9028	+11.8	−29.1	− 4.1
4399	−19.7	+ 4.6	− 5.5	9072	+ 8.8	−14.6	−11.6
4408	+26.0	−17.2	+ 7.4				

[a] The first column contains the number of the star in the *Yale Catalogue of Bright Stars*. The other three columns give the U, V, and W components of the star's velocity, in km/s. The data are from unpublished work of the author with Prof. A. Blaauw, Groningen.

THE DISTRIBUTION OF THE RESIDUAL VELOCITIES:
A NUMERICAL METHOD

1. In the preceding exercise we have investigated the distribution of the residual velocities of a group of stars by a graphic method. Now we shall resume this investigation but instead apply a numerical method which essentially represents a translation of the previous one into mathematical terms.

Let us begin by recalling that, according to the geometric method adopted, the coordinate of a representative point with respect to any axis is taken to be equal to the component, along the same axis, of the velocity to which this point corresponds.

Consider now the point U_i, V_i, W_i, representing the velocity of the i-th star referred to the original system of axes. The coordinates of the same point with respect to axes parallel to the original axes, but shifted into the center of the distribution, i.e. into the point $\bar{U}$, $\bar{V}$, $\bar{W}$, will be

$$U_i - \bar{U}, \quad V_i - \bar{V}, \quad W_i - \bar{W}.$$

Take now an arbitrary axis x drawn through the center of the distribution and let its zero point coincide with the center of the distribution. Let, further, l, m, and n, be the direction cosines of this new axis with respect to the shifted ones. Then, according to the formula for a rotated axis, the coordinate U_{xi} of the point i, with respect to the x-axis is given by

$$U_{xi} = l(U_i - \bar{U}) + m(V_i - \bar{V}) + n(W_i - \bar{W}). \qquad \text{(V.1)}$$

Let us again adopt, as the measure of the scatter of the components U_{xi}, the quantity σ_x defined by

$$\sigma_x^2 = \frac{1}{n}\sum_{i=1}^{n} U_{xi}^2 = \frac{1}{n}\sum_{i=1}^{n}[l(U_i - \bar{U}) + m(V_i - \bar{V}) + n(W_i - \bar{W})]^2. \qquad \text{(V.2)}$$

Notice that σ_x^2 represents, in some way, a generalisation of the mean square deviation currently used to characterise the distribution of the values of a random variable around its mean, and realise that this is just one possibility to arrive at a quantitative estimate of the distribution of the representative points. It is beyond the scope of this exercise to justify this choice, or to discuss its shortcomings and consider other possibilities. The interested reader will find, e.g. in the monographs by Ogorodnikov [1], and Trumpler and Weaver [2], an account of these questions adapted to his needs. All we can do here is to derive formulae which represent a numerical counterpart of the graphic method previously used.

The value of σ_x we effectively can find for a given distribution depends on the direction of the x-axis adopted. This is clearly shown by Equation (V.2) according to which, for a given set of values of U_i, V_i, and W_i, $(i=1, 2, ..., n)$ as well as $\bar{U}$, $\bar{V}$, $\bar{W}$, the quantity σ_x^2 is a function of the direction cosines l, m, and n. Now we have seen that in certain directions, which correspond to what we have called the axes of the distribution, σ_x^2 attains extreme values. In Exercise IV, Section 5, we have agreed to characterise quantitatively the distribution of the residual velocities by specifying, first, the directions in which σ_x^2 attains the extreme values, and, second, by the values attained. In our numerical analysis we will adopt this same principle.

2. The problem to be solved can be stated as follows: Find the extreme values of the function

$$F(l, m, n) = \frac{1}{n} \sum_{i=1}^{n} [l(U_i - \bar{U}) + m(V_i - \bar{V}) + n(W_i - \bar{W})]^2, \qquad (V.3)$$

the variables l, m, and n, being subject to the condition

$$f(l, m, n) = l^2 + m^2 + n^2 - 1 = 0. \qquad (V.4)$$

The relation (V.4) follows from the fact that the variables l, m, and n, are the direction cosines of an axis and that, therefore, the sum of their squares must be equal to unity. Our problem is one of extreme values with side conditions. It is almost identical to that discussed in Problem I.1. Nevertheless, for sake of clarity and completeness, its solution will be considered anew.

Due to the existence of a relation between the variables, in the present case of the relation (V.4), the extreme values cannot be found by simply equating to zero the partial derivatives of $F(l, m, n)$. Making use of Lagrange's method of indeterminate multipliers (see any textbook of mathematical analysis) the solution can be obtained as follows.

Multiply the equation of condition (V.4) by the (provisionally) indeterminate multiplier $-\lambda$, and add to (V.3). This gives

$$\bar{F}(l, m, n) = F(l, m, n) - \lambda f(l, m, n) \qquad (V.5)$$

or explicitly

$$\bar{F}(l, m, n) = (\mu_{11} - \lambda) l^2 + (\mu_{22} - \lambda) m^2 + (\mu_{33} - \lambda) n^2$$
$$+ 2\mu_{12}lm + 2\mu_{13}ln + 2\mu_{23}mn + \lambda \qquad (V.5a)$$

where we have put

$$\mu_{11} = \frac{1}{n} \sum_{i=1}^{n} (U_i - \bar{U})^2 \quad \mu_{12} = \frac{1}{n} \sum_{i=1}^{n} (U_i - \bar{U})(V_i - \bar{V}) \quad \mu_{13} = \frac{1}{n} \sum_{i=1}^{n} (U_i - \bar{U})(W_i - \bar{W})$$

$$\mu_{22} = \frac{1}{n} \sum_{i=1}^{n} (V_i - \bar{V})^2 \qquad \mu_{23} = \frac{1}{n} \sum_{i=1}^{n} (V_i - \bar{V})(W_i - \bar{W})$$

$$\mu_{33} = \frac{1}{n} \sum_{i=1}^{n} (W_i - \bar{W})^2. \qquad (V.6)$$

By developing the sums and making use of the definition of the mean values we also can write

$$\mu_{11} = \frac{1}{n} \sum_{i=1}^{n} U_i^2 - (\bar{U})^2 \qquad \mu_{12} = \frac{1}{n} \sum_{i=1}^{n} U_i V_i - \bar{U} \cdot \bar{V} \qquad \mu_{13} = \frac{1}{n} \sum_{i=1}^{n} U_i W_i - \bar{U} \bar{W}$$

$$\mu_{22} = \frac{1}{n} \sum_{i=1}^{n} V_i^2 - (\bar{V})^2 \qquad \mu_{23} = \frac{1}{n} \sum_{i=1}^{n} V_i W_i - \bar{V} \bar{W}$$

$$\mu_{33} = \frac{1}{n} \sum_{i=1}^{n} W_i^2 - (\bar{W})^2. \tag{V.7}$$

The necessary conditions for an extremum are now

$$\frac{\partial F}{\partial l} = 0, \qquad \frac{\partial F}{\partial m} = 0, \qquad \frac{\partial F}{\partial n} = 0 \tag{V.8}$$

or

$$\begin{aligned}
(\mu_{11} - \lambda) l + & \quad \mu_{12} m + & \quad \mu_{13} n = 0 \\
\mu_{12} l + & (\mu_{22} - \lambda) m + & \quad \mu_{23} n = 0 \\
\mu_{13} l + & \quad \mu_{23} m + & (\mu_{33} - \lambda) n = 0.
\end{aligned} \tag{V.9}$$

The Equations (V.9) represent a system of homogeneous linear equations. Apart from the trivial solution $l = m = n = 0$ they can be satisfied by a set of values of l, m, and n, only if the determinant of the coefficients is equal to zero:

$$D(\lambda) = \begin{vmatrix} \mu_{11} - \lambda & \mu_{12} & \mu_{13} \\ \mu_{12} & \mu_{22} - \lambda & \mu_{23} \\ \mu_{13} & \mu_{23} & \mu_{33} - \lambda \end{vmatrix} = 0. \tag{V.10}$$

We shall accept without proof that this cubic for λ has three real and different roots $\lambda = \lambda_i$ ($i = 1, 2, 3$). By substituting successively each of them into (V.9) we will get three systems of linear equations from which the corresponding direction cosines l_i, m_i, and n_i, can be found and so the direction of the axes of the distribution determined. Finally, by substituting these values of the direction cosines into (V.2) we can compute the corresponding dispersions, which completely solves the problem.

3. In our account of the method of solution we have so far followed the same lines as in Problem I.1. However as now we are interested in a numerical solution of our present problem, some important details must be considered more closely. Let us begin with the computation of the roots λ_i.

Develop the determinant (V.10). This gives

$$- D(\lambda) = g(\lambda) = \lambda^3 + k_1 \lambda^2 + k_2 \lambda + k_3 = 0 \tag{V.11}$$

where the coefficients k_1, k_2, and k_3 are given by

$$\begin{aligned}
k_1 &= - (\mu_{11} + \mu_{22} + \mu_{33}), \\
k_2 &= \mu_{11}\mu_{22} + \mu_{11}\mu_{33} + \mu_{22}\mu_{33} - (\mu_{12}^2 + \mu_{13}^2 + \mu_{23}^2), \\
k_3 &= \mu_{12}^2\mu_{33} + \mu_{13}^2\mu_{22} + \mu_{23}^2\mu_{11} - \mu_{11}\mu_{22}\mu_{33} - 2\mu_{12}\mu_{13}\mu_{23}.
\end{aligned} \tag{V.12}$$

Now, if nothing is known about the values of the roots of (V.11) one may be inclined to make use of the formulae giving the exact solution, which can be found in any text-book on algebra. In practice however it will be found more profitable first to derive in some way approximate values of the roots which then will be used to start an iterative process leading to values of the roots as exact as needed. As the approximate values one can use the μ's standing on the diagonal of the determinant, i.e. μ_{11}, μ_{22}, and μ_{33}, respectively. The iterative process to be used is that known as the Newton-Raphson method. Let $\lambda_i^{(1)}$ be an approximative value of the root λ_i. Then a better one will be given by $\lambda_i^{(2)} = \lambda_i^{(1)} + \Delta\lambda_i^{(1)}$, where

$$\Delta\lambda_i^{(1)} = - g\left(\lambda_i^{(1)}\right)/g'\left(\lambda_i^{(1)}\right). \tag{V.13}$$

Using $\lambda_i^{(2)}$ compute $\Delta\lambda_i^{(2)} = -g\left(\lambda_i^{(2)}\right)/g'\left(\lambda_i^2\right)$, which in general will be smaller in absolute value than the preceding correction, so that $\lambda_i^{(3)} = \lambda_i^{(2)} + \Delta\lambda_i^{(2)}$ can be considered as an even better approximation. Use again $\lambda_i^{(3)}$ to find the following approximation $\lambda_i^{(4)} = \lambda_i^{(3)} + \Delta\lambda_i^{(3)}$, continuing in this way until the correction $\Delta\lambda_i^{(n)}$ becomes smaller than one half unit of the last place retained.

Having determined all the three roots λ_1, λ_2, and λ_3, with the desired precision, we can proceed to the solution of the three systems of linear equations for the direction cosines. Let λ_i be one of the three roots. Then

$$\begin{aligned}(\mu_{11} - \lambda_i)\, l_i + & \quad \mu_{12} m_i + & \mu_{13} n_i = 0 \\ \mu_{12} l_i + (\mu_{22} - \lambda_i)\, m_i + & \quad \mu_{23} n_i = 0 \\ \mu_{13} l_i + & \quad \mu_{23} m_i + (\mu_{33} - \lambda_i)\, n_i = 0\end{aligned} \tag{V.14}$$

will be the system of homogeneous linear equations for the corresponding values l_i, m_i, and n_i, of the direction cosines.

First consider the last two equations. Divide by n_i taking provisionally l_i/n_i and m_i/n_i as the unknowns. Then it is easily found that

$$l_i = \frac{G_{11}^i}{G_{13}^i}\, n_i, \qquad m_i = \frac{G_{12}^i}{G_{13}^i}\, n_i, \qquad n_i = n_i \tag{V.15}$$

where

$$G_{11}^i = \begin{vmatrix} \mu_{22} - \lambda_i & \mu_{23} \\ \mu_{23} & \mu_{33} - \lambda_i \end{vmatrix}, \quad G_{12}^i = - \begin{vmatrix} \mu_{12} & \mu_{23} \\ \mu_{13} & \mu_{33} - \lambda_i \end{vmatrix},$$

$$G_{13}^i = \begin{vmatrix} \mu_{12} & \mu_{22} - \lambda_i \\ \mu_{13} & \mu_{23} \end{vmatrix}. \tag{V.16}$$

Notice that G_{11}^i, G_{12}^i, and G_{13}^i, are the subdeterminants corresponding to the first row of the determinant (V.10), where λ_i has been substituted for λ.

Now the direction cosines are subject to the condition (V.4). Substituting the values of l_i, m_i, and n_i (V.4) one easily finds

$$l_i = G_{11}^i/R_i, \qquad m_i = G_{12}^i/R_i, \qquad n_i = G_{13}^i/R_i. \tag{V.17}$$

Where

$$R_i = (G_{11}^i)^2 + (G_{12}^i)^2 + (G_{13}^i)^2. \tag{V.18}$$

Remember now that the direction cosines of any direction with respect to a given system of axes are equal to the components, along the same axes, of the unit vector which corresponds to this direction. Notice further that the U–V–W-system of axes used so far differs from the system of galactic coordinates only by the direction of its first (i.e. U-) axis, which is opposite to that of the first axis of the usual galactic coordinate system. If therefore l_i, m_i, n_i, are the direction cosines of any direction with respect to the axes U, V, and W, respectively, then the direction cosines of the same direction referred to the axes of the usual galactic coordinate system will be $-l_i$, m_i and n_i.

Denote by L_i and B_i respectively the galactic longitude and the galactic latitude of the directions which correspond to extreme values of the dispersion σ_x^2. Then obviously we will have

$$\operatorname{tg} L_i = -\,m_i/l_i, \qquad \sin B_i = n_i, \quad i = 1, 2, 3. \tag{V.19}$$

Now it can easily be shown that the directions just referred to are perpendicular to each other. For that purpose write the two systems of linear Equations (V.14) which correspond to two different solutions as follows:

$$
\begin{aligned}
\lambda_i l_i &= \mu_{11} l_i + \mu_{12} m_i + \mu_{13} n_i & \lambda_j l_j &= \mu_{11} l_j + \mu_{12} m_j + \mu_{13} n_j \\
\lambda_i m_i &= \mu_{12} l_i + \mu_{22} m_i + \mu_{23} n_i & \lambda_j m_j &= \mu_{12} l_j + \mu_{22} m_j + \mu_{23} n_j \\
\lambda_i n_i &= \mu_{13} l_i + \mu_{23} m_i + \mu_{33} n_i & \lambda_j n_j &= \mu_{13} l_j + \mu_{23} m_j + \mu_{33} n_j.
\end{aligned}
\tag{V.20}
$$

Multiply the equations of the left group respectively by l_j, m_j, and n_j, and add. Similarly, multiply the equations of the right group by l_i, m_i, and n_i, and add. Take the difference of the two sums. Then due to the symmetry of the matrix of the coefficients μ_{pq} we get:

$$(\lambda_i - \lambda_j)\,(l_i l_j + m_i m_j + n_i n_j) = 0. \tag{V.21}$$

As, however, all the roots λ_i are different, this necessarily means that

$$l_i l_j + m_i m_j + n_i n_j = 0 \tag{V.22}$$

which is the condition of orthogonality of the directions defined by the direction cosines l_i, m_i, n_i, and l_j, m_j, n_j.

The relations (V.4) and (V.22) can be used to check the solutions.

In order to find the extreme values of the dispersion perform the operations indicated in (V.2) and make use of the notations (V.7). This gives

$$
\begin{aligned}
\sigma_x^2 = \mu_{11} l^2 \;&+ \mu_{12} lm \;+ \mu_{13} ln \\
+\,&\mu_{12} ml + \mu_{22} m^2 + \mu_{23} mn \\
+\,&\mu_{13} nl \;+ \mu_{23} nm + \mu_{33} n^2,
\end{aligned}
\tag{V.23}
$$

or

$$
\begin{aligned}
\sigma_x^2 = l\,(\mu_{11} l \;&+ \mu_{12} m + \mu_{13} n) \\
+\, m\,(\mu_{12} l \;&+ \mu_{22} m + \mu_{23} n) \\
+\, n\,(\mu_{13} l \;&+ \mu_{23} m + \mu_{33} n).
\end{aligned}
\tag{V.24}
$$

Making use of equations of the left group (V.20) and taking into account (V.4) one easily finds that the dispersion σ_i corresponding to the direction $l_i, m_i, n_i, (i=1, 2, 3)$ is equal to $\pm\sqrt{\lambda_i}$. This completely solves the problem.

4. From a mathematical standpoint our foregoing developments represent the determination of the proper vectors and of the proper values of the symmetric tensor

$$\begin{pmatrix} \mu_{11} & \mu_{12} & \mu_{13} \\ \mu_{12} & \mu_{22} & \mu_{33} \\ \mu_{13} & \mu_{23} & \mu_{33} \end{pmatrix}. \tag{V.25}$$

That (V.25) is indeed a tensor can be shown by making use of the transformation properties of the tensors and by proving that the quantities μ_{pq} conform to the formulae for tensor components. When this has been done, the problem which we have discussed can be stated as one of determining the proper values and proper vectors. The solution can, then, immediately be written down. However the tensor calculus is usually considered in a quite different context. Therefore we have refrained from such an approach and, instead, preferred to derive the necessary formulae from the very beginning.

Concluding we may say that we have characterised the distribution of the residual velocities by a symmetric tensor. Its components μ_{pq} are the so-called central second order moments of the distribution.

It is beyond the scope of this exercise to discuss the relation between the quantity which we have used to characterise the distribution of the residual velocities and the parameters of the velocity distribution function. As far as this question is concerned the reader is referred to the textbook by D. Mihalas and P. McRae Routly [3], and especially to the monographs by Ogorodnikov, and Trumpler and Weaver, already cited, which also give reference to earlier work in this field. A short but very clear and useful comment will be found in a paper by J. B. Alexander [4]. Here we can only mention that in our notations, the equation of the so-called velocity ellipsoid will be

$$\frac{(U - \bar{U})^2}{\sigma_1^2} + \frac{(V - \bar{V})^2}{\sigma_2^2} + \frac{(W - \bar{W})^2}{\sigma_3^2} = 1. \tag{V.26}$$

The method explained here is, in essence, that of C. V. L. Charlier [5]. It has been extensively used by Charlier himself as well as by his pupils. Our account has been inspired by that given by Ogorodnikov [1].

Let us finally remark that for an analysis of the distribution of the residual velocities one can also make use of the radial velocities or of the proper motions.

It is proposed to the reader that he should derive, by the method shown above, the direction of the axes and the corresponding dispersions of the distribution of the residual velocities of the stars listed in Table IV.1 in Exercise IV, p. 64. In view of the volume of computational work involved the reader is expressly advised to make use of an appropriate computing form. The most tedious part of the work is the computation of the moments μ_{pq} for which some type of printing calculator should be used, as it permits one to check easily the values obtained. If he has no such calculator at his disposal, the reader is advised first to compute the sums (V.9) for small groups each containing e.g. ten stars (in our case the last will obviously contain only five), admitting that the results are correct if by repeated calculation the same values are obtained. The values of the moments can then be computed by adding the partial results and by dividing by the number of stars (n). Such calculations can readily be made with the aid of any standard desk calculator. Obviously the work will be greatly alleviated if use can be made of a programmable electronic desk calculator. For the reader's convenience we give on the following pages the results at some of the critical points and the final results.

The interested reader is referred to [2] and [4] for other examples of such calculations.

From Table VI.1, p. 74 we find

$$\sum U = + \quad 731.4 \qquad \sum V = - \quad 700.8 \qquad \sum W = - \quad 537.4$$
$$\sum U^2 = \quad 40346.20 \qquad \sum V^2 = \quad 21493.06 \qquad \sum W^2 = \quad 11853.74$$
$$\sum UV = - 13832.49 \qquad \sum VW = + \quad 4801.74 \qquad \sum WU = - \quad 6167.04 .$$

By Equation (IV.4), p. 52 we have

$$\bar{U} = + 9.75, \qquad \bar{V} = - 9.34, \qquad \bar{W} = - 7.17, \quad \text{km/s} .$$

By Equations (IV.5) and (IV.6), p. 53 we have for the elements of the solar motion relative to the 75 early F-type Main Sequence stars,

$$\text{tg } l_A = + 0.958 \qquad \sin b_A = + 0.469 ,$$
$$l_A = 43°.8 , \qquad b_A = + 28°.0 , \qquad S = 15.3 \text{ km/s} .$$

Having computed, according to Equations (V.7) the quantities μ_{pq} we get the cubic for λ which is

$$\begin{vmatrix} + 442.848 - \lambda & - 93.311 & - 12.351 \\ - 93.311 & + 199.264 - \lambda & - 2.930 \\ - 12.351 & - 2.930 & + 106.708 - \lambda \end{vmatrix} = 0 .$$

For the roots λ_i and the corresponding main dispersions we have

$$\lambda_1 = 474.80 \qquad \lambda_2 = 168.36 \qquad \lambda_3 = 105.66$$
$$\sigma_1 = \pm 21.79 \qquad \sigma_2 = \pm 12.97 \qquad \sigma_3 = \pm 10.27, \text{ km/s} .$$

Making use of the formulae (V16), (V17), and (V.18) we get for the direction cosines of the main axes

$$l_1 = + 0.9469 \qquad m_1 = - 0.3203 \qquad n_1 = - 0.0292$$
$$l_2 = - 0.3156 \qquad m_2 = - 0.9427 \qquad n_2 = + 0.1080$$
$$l_3 = + 0.0624 \qquad m_3 = + 0.0933 \qquad n_3 = + 0.9937 .$$

Verify that the conditions of orthonormality (V.4) and (V.22), pp. 66 and 69 are satisfied.

The direction of the main axes as given by the galactic coordinates L_i and B_i follow from the formulae (V.19), p. 69. We have:

$$\operatorname{tg} L_1 = + 0.3383 \qquad L_1 = 18^\circ\!.7 \qquad B_1 = - 1^\circ\!.7$$
$$\operatorname{tg} L_2 = - 2.987 \qquad L_2 = 288^\circ\!.5 \qquad B_2 = + 6^\circ\!.2$$
$$\operatorname{tg} L_3 = - 1.496 \qquad L_3 = 123^\circ\!.8 \qquad B_3 = + 83^\circ\!.5 .$$

References

[1] Ogorodnikov, K. F.: 1965, *Dynamics of Stellar Systems*, Pergamon Press, Oxford.
[2] Trumpler, R. J. and Weaver, H. F.: 1953, *Statistical Astronomy*, University of California Press, Berkeley. Reprint: Dover Publications, New York, 1962.
[3] Mihalas, D., with the collaboration of McRae Routly, P.: 1968, *Galactic Astronomy*, Freeman and Co., San Francisco.
[4] Alexander, J. B.: 1958, *Monthly Notices Roy. Astron. Soc.* **118**, 161.
[5] Charlier, C. V. L.: 1926, *The Motion and the Distribution of the Stars*, University of California Press, Berkeley.

THE ASYMMETRY OF STELLAR MOTIONS

1. In the preceding exercises we have mainly been interested in determining some specific quantity. The present exercise will be devoted to a rather qualitative investigation aimed at illustrating certain aspects of the so-called asymmetry of the stellar motions. At the same time we shall very briefly summarise the results of the many determinations of the solar apex with respect to different types of stars and other objects, as well as of the parameters of their velocity distribution. The reader is expressly advised to consult, for more details, the most recent and comprehensive account on this subject given by J. Delhaye in an article already cited [1].

Let us begin with some definitions taken from Delhaye.

The Standard Solar Motion is the solar motion with respect to the stars which form the majority in the general catalogues of radial velocities and proper motions (A to G main sequence stars, giants and supergiants). It is given by the velocity vector which has the following components:

$$U_s = -10.4, \qquad V_s = +14.8, \qquad W_s = +7.3, \quad \text{km/s}.$$

Let us recall that the first (or U-) axis of the system of axes used points towards $l = 180°$, $b = 0°$, the other two being identical to the second and third axis of the usual (new) galactic coordinate system.

The same vector can also be defined by its absolute value and direction, but now with respect to the usual (new) galactic coordinate axes:

$$S_s = 19.5 \text{ km/s}; \qquad l_s = 56°, \qquad b_s = +23°.$$

The Basic Solar Motion is defined by the most frequently occurring velocities in the solar neighborhood. It may also be said that it is the velocity of the Sun with respect to the majority of the stars in our vicinity. The vector of the Basic Solar Velocity is given by the following components:

$$U_B = -9, \qquad V_B = +11, \qquad W_B = +6, \quad \text{km/s}$$

or, as above, by its absolute value (in km/s) and the galactic coordinates of the point towards which it is directed:

$$S_B = 15.4 \text{ km/s}; \qquad l_B = 51°, \qquad b_B = +23°.$$

Now if you inspect the long list of apex determinations with respect to different classes of stars and other objects given by Delhaye (Table 1, on pp. 64–67 in his article), you will find that for some classes the absolute value of the solar velocity as well as its direction differ considerably from the values given above. The following excerpt from Delhaye's Table illustrates this fact well.

TABLE VI.1[a]

Solar motion relative to some classes of objects

	S	l_A	b_A
Planetary nebulae	31	74°	14°
Subgiants	36.4	75	15
White dwarfs	38	81	12
RR Lyrae variables (Period $< 0.45^d$)	56	32	4
Subdwarfs	149	83	2
Globular clusters	182	87	2
RR Lyrae variables (Period $> 0.45^d$)	218	90	6

[a] Data from Delhaye, J.: 1965, in A. Blaauw and M. Schmidt (eds.) *Galactic Structure*, University Press, Chicago, pp. 64–67.

In order to get a better insight into the real meaning of these data we shall display them on a graph. Before doing this let us, however, make two remarks. First, notice that with one or two exceptions, the galactic latitudes of all the apexes are very low. In other words all the velocity vectors are practically in the galactic plane. This means that our graphic representation can be limited to the UV-plane. The second remark concerns the choice of the point of reference i.e. of the zero-point of the system of UV-axes. It is obvious that in an investigation of the motions of different classes of objects in our stellar system the motion of the Sun itself cannot be regarded as an especially important datum. However the data given in our Table VI.1 can be used to investigate the relative motions of the different classes of stars to which they pertain. In order to achieve this we shall represent in the UV-diagram their velocities relative to the Sun, and not the Sun's velocity itself.

Let now S, l_A, b_A, be the elements of the solar motion relative to a class of objects. Then the components of the solar velocity along the axes of the galactic coordinate system will respectively be equal to

$$u = S \cos b_A \cos l_A, \qquad v = S \cos b_A \sin l_A, \qquad w = S \sin b_A, \qquad \text{(VI.1)}$$

whereas the components of the velocity of the class considered, relative to the Sun will obviously be

$$-u, \quad -v, \quad -w.$$

Bearing in mind that the U-axis has a direction opposite to that of the u-axis we finally get

$$U = S \cos b_A \cos l_A, \qquad V = -S \cos b_A \sin l_A, \qquad W = -S \sin b_A. \qquad \text{(VI.2)}$$

Using the data from Table VI.1 compute the components U and V of the velocity of each of the classes of objects given there, and make the corresponding UV-diagram.

Add one point more for the velocity of the bulk of stars in the solar neighborhood (basic solar motion). Although it does not pertain to a physically well defined class, it can nevertheless be considered as a representative value for the motion of the so-called common stars. Figure VI.1 represents schematically the sort of diagram you will get.

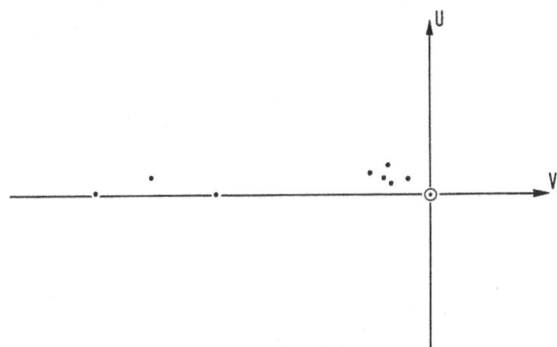

Fig. VI.1.

As you will see from your diagram all the representative points are disposed along an axis which is called the asymmetry axis. This axis is very nearly perpendicular to the direction towards the galactic center. Moreover, as it can be seen from Table VI.1, the dispersions of the velocity components in general, and those of the U components in particular, increase from right to left along the asymmetry axis, being of the order of some tens of km/s for the subgiants, and attaining values far above 100 km/s for the RR Lyrae stars with periods longer than 0.45 d.

The phenomenon disclosed is the so-called asymmetric drift found and investigated in the early twenties by G. Strömberg [2]. A more recent study covering this as well as other related subjects has been made by P. P. Parenago in an important paper [3].

Summarising his investigations G. Strömberg says: "The general result of this study is that the Sun's motion is not a constant vector but changes greatly with the class of object to which the motion is referred." Moreover he finds "that a very definite relation exists between group motion and internal motion for different classes of objects", which can be expressed in the form of a linear equation relating the absolute value of the velocity S to the square of a dispersion which roughly corresponds to our σ_1^2. It is beyond the scope of this exercise to discuss this question. Let us only remark that the existence of such a relation can be derived from the so-called hydro-dynamical equations of a stationary Galaxy (B. Lindblad). However from the beginning, doubts have been expressed as to the general validity of Strömberg's relation. For more details the reader is referred to Delhaye's article.

Notice finally that, if we arrange the different classes of objects according to the position of the corresponding representative point on the asymmetry axis (counted from left to right), then they will at the same time be arranged in order of increasing concentration towards the galactic plane or, what amounts to the same, in order of increasing flattening. See e.g. Table 3 in A. Blaauw's article on stellar population [4].

2. In a paper published in 1927 B. Lindblad [5] proposed the following interpretation of the asymmetry of stellar motions:

We assume that the stellar system may be divided up into a series of subsystems having rotational symmetry around one and the same axis, with different speeds of rotation at the same distance from the axis and consequently having different degrees of flattening. The inner, most flattened, systems have a high star density, though decreasing with decreasing speed of rotation; the systems with the highest speed of rotation are assumed to form the Milky Way clouds. In the extreme outer systems the space density of the individuals, stars or globular clusters, is relatively low. The latter sub-systems show, on account of their low speed of rotation, a strong asymmetrical drift in velocity nearly at a right angle to the radius of the big system, when the velocity of their members are measured from a star, like our Sun, moving as a member of a Milky Way cloud.

In our present terminology members of the Milky Way clouds are, generally speaking, Population I stars; 'a strong asymmetrical drift' means that the representative point on the UV-diagram is far to the left on the asymmetry axis, whereas the radius of the big system corresponds to the direction towards the center of the Galaxy. Or, quoting once more from B. Lindblad [6], "The phenomenon of the asymmetrical drift depends on the difference of rotation of the different sub-systems". The objects we now call extreme Population I have the largest rotational velocity. In other words, the stars which on account of their velocity relative to the Sun we call high velocity stars are in fact slow-moving members of our system, lagging behind the faster ones, to which our Sun belongs.

In the second paper quoted Lindblad says further that "A confirmation of the ideas just developed is obtained from Oort's result of the existence of a velocity limit above which the directions of stellar motions show an absolute avoidance of one hemisphere". Assuming correctly that the Sun belongs to a very flattened sub-system, but supposing (what is now known to be wrong) that it is close to the very edge of this sub-system, Lindblad proceeds as follows:

At the edge of a very flattened spheroid the velocity of motion in circular orbits is very near the velocity of escape. Consequently, in the case of the Milky Way clouds we can have only a small relative velocity in direction of the speed of rotation, if the velocity of the star is supposed to stay below the velocity of escape.

The relative velocity referred to is the velocity with respect to a point moving in a circular orbit. The velocity of escape is the velocity required to escape from the system by overcoming the attractional force.

More direct proofs of the theory of galactic rotation will be considered in the following Exercise. At this place let us quote some of the most important ideas put forward by J. H. Oort in this context, and thereafter propose to the reader to repeat in a simplified and abridged form some of his investigations.

3. In a paper published in 1928 [7], in which he resumes his earlier investigations referred to by Lindblad, Oort refutes, on ground of observational evidence (density distribution of the stars) Lindblad's idea that the Sun may be at the very edge of our stellar system. He proposes to reconsider in the present article some of the consequences of the rotation hypothesis, in particular those with respect to the distribution of the peculiar velocities of the stars in general, and to the motion of the high velocity stars ...

It is very unlikely that the galactic system as a whole should contain a considerable number of stars whose velocities exceed the velocity of escape. Let us call the velocity of escape from a point near the Sun V_e and let Θ_0 represent the linear velocity of rotation of, say, the apparently brightest stars. Let further the velocities of the stars be counted from this rotational velocity as origin. It is then clear that, if there are any velocities larger than $V_e - \Theta_0$ these velocities must not be directed to a region surrounding the direction of rotational velocity: they should avoid the area round 54° galactic longitude [which in the system of galactic coordinates used by Oort corresponds to the direction at a right angle to the direction towards the galactic center].

Now this is exactly the state of things that has been shown to exist with the so-called high velocity stars. A large area is avoided by stars with velocities higher than 65 km/s and it is clear at a glance that the center of this area lies on the galactic circle [that is on the galactic equator] in the neighborhood of the predicted point.

Let us add some comment. Peculiar velocity here means velocity referred to the centroid of the apparently brightest stars. These can be identified with the stars forming the majority in our catalogues of radial velocities and proper motions according to Delhaye's definition. The quantity Θ_0 which we shall always consider as a vector $\mathbf{\Theta}_0$ is the velocity due to the rotation around the galactic axis. It is referred to a frame having its origin in the center of gravity of the galactic system, with axes parallel to the usual U, V, and W axes, at a given moment of time. It is assumed that $\mathbf{\Theta}_0$ is in the galactic plane and has a direction perpendicular to the direction towards the galactic center. Therefore $\mathbf{\Theta}_0$ is parallel to the V-axis.

Now notice that in any point in our Galaxy there must exist, for all its permanent members, an upper limit to the velocity as measured relative to the fixed frame. Its absolute value is determined by the obvious condition that the kinetic energy to which it corresponds is equal to the work to be done against the attractive force of the Galaxy in order to bring the star from its actual position at rest in infinity. It is this velocity which is called the velocity of escape. Obviously V_e will change from one point to another in our stellar system. But in a first approximation it can be assumed that, in a limited volume of space, the variation of V_e can be neglected.

Let now U, V, W, be the components of the peculiar velocity of a star. The components of its velocity relative to the frame at rest will be, with $\Theta_0 = |\mathbf{\Theta}_0|$

$$U, \quad V + \Theta_0, \quad W.$$

Our condition is that for all permanent members of our stellar system

$$[U^2 + (V + \Theta_0)^2 + W^2] < V_e^2. \tag{VI.3}$$

Consider now the equation

$$U^2 + (V + \Theta_0)^2 + W^2 = V_e^2. \tag{VI.4}$$

According to our geometric method this is the equation of a sphere of radius V_e having its center on the V-axis, and the condition (VI.3) can be interpreted by saying that the representative points which correspond to the velocities of the permanent members of the Galaxy in the solar neighborhood must be contained in this sphere.

The relation (VI.3) has consequences which can be verified. In fact put $V_e - \Theta_0 = V_{pec}$ and consider the stars with peculiar velocities which are, by absolute value, smaller than, or equal to, V_{pec}. The end points of the corresponding velocity vectors will lie within the sphere of radius V_{pec} drawn around the end point of Θ_0 and tangent to the sphere (VI.4), see Figure VI.2a. Therefore, regardless of the direction, all of them will also be contained in the larger sphere. In other words, as far as the condition (VI.3) is concerned, all directions are permitted. If we make the UV-diagram of these peculiar velocities, then the representative points will populate the small circle of radius V_{pec} tangent to the larger circle of radius V_e. They will be distributed symmetrically about the origin which corresponds to the end point of the velocity vector Θ_0, see Figure VI.2b.

However, as soon as we consider velocities larger by absolute value than V_{pec}, say those between V_{pec} and some $V' > V_{pec}$, the angular distribution of the velocity

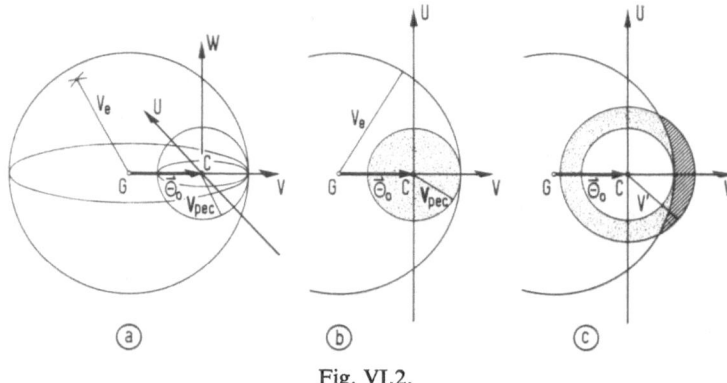

Fig. VI.2.

vectors will lose its isotropic character. For there will now be, in the spherical shell between the two concentric spheres with radii V_{pec} and V' respectively, a region lying outside the large sphere of allowed velocities (VI.4). As seen from the corresponding UV-diagram the distribution of the representative points will no more be symmetric about the origin, see Figure VI.2c.

4. It is proposed that the reader should verify the effect predicted. As sources of data on stellar velocities he can use:

Eggen, O. J.: 1962, Space Velocity Vectors for 3483 Stars, Royal Observatory Bulletins, No. 51, London.
Eggen, O. J.: 1964, A Catalogue of High-Velocity Stars, Royal Observatory Bulletins, No. 84, London.

We propose that the reader should proceed as follows:

I. To pick out, from the first of Eggen's Catalogues, some 20 stars with space velocities lower than 50 km/s, regardless of other characteristics, and spread over the whole

Catalogue. As he will see, it is not difficult to get, by simple inspection of the velocity components, a fair estimate of absolute value of the space velocity.

Make now the UV-diagram as in Exercise IV, but adopt a smaller scale (e.g. 1 mm for 10 km/s). As the values of the components given in the Catalogue are heliocentric, the center of the distribution will be somewhat displaced with respect to the origin of your diagram. Make, further, a second diagram in the same way, but taking stars with space velocities between 50 km/s and 100 km/s, and, finally a third one for stars with space velocities above 100 km/s. As you will see, the cloud of representative points will grow in both directions along the U-axis. However it will grow only in the negative direction of the V-axis, so that the region around galactic longitude $l = 90°$ (which corresponds to the positive direction of the V-axis) will in fact be avoided. It is in principle in this way that Oort has derived the value of 63 km/sec for V_{pec}.

The same effect is present in the velocity distribution of any physically homogeneous class of stars. A list of such stars is given in Table VI.2. The data have been taken from Eggen's catalogue of space velocity vectors.

Fig. VI.2b and c, shows schematically the kind of diagram you should obtain. II. Instead of the space velocities one can use the radial velocities, as was done by Oort [7]. For that purpose take in the *Yale Catalogue of Bright Stars* [8] all the stars with radial velocities above, e.g., 80 km/s, and proper motions $\mu'' < 0.3''$. This last restriction means that we wish to eliminate stars with large transversal velocities, so that the radial velocity can be considered as an acceptable substitute for the space velocity. Taking the Sun as the origin, make a diagram of the angular distribution of the radial velocities by drawing, in the UV-plane, vectors which correspond to the radial velocity (not to its projection) and which shall be directed towards the star if the star is moving away from the Sun (positive radial velocity), but which, for stars with negative radial velocities, shall be directed towards the galactic longitude $l_{star} + 180°$ see Figure VI.3a. The galactic longitudes of the stars are given in the *Yale Catalogue*. As the catalogue values of the radial velocities are heliocentric, it would be, strictly

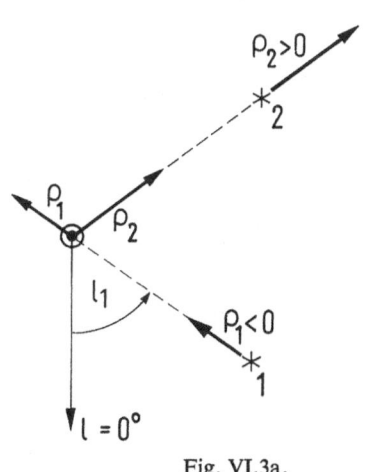

Fig. VI.3a.

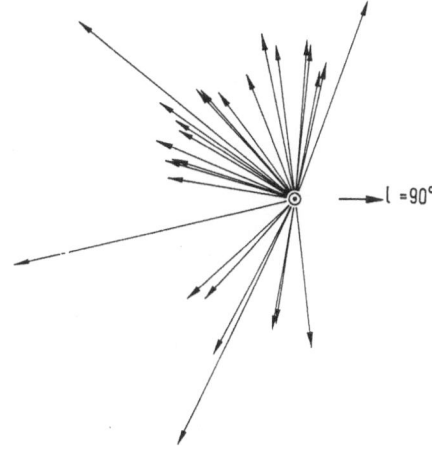

Fig. VI.3b.

speaking, necessary to correct them for the effect of solar motion. In order to avoid the rather tedious reductions, the lower limit of the radial velocity has been chosen in such a way, that even with uncorrected radial velocities the asymmetry will be clearly exhibited. Figure VI.3b represents schematically the type of diagram you will get. A list of stars with radial velocities above 80 km/s is given in Table VI.3. The data have been taken from the *Yale Catalogue of Bright Stars*.

Most of the necessary data have already been collected in the accompanying tables. Nevertheless the reader who has at his disposal the original catalogues is expressly advised to prepare himself the lists, and to read the Introductions to the Catalogues, where he will find valuable information.

References

[1] Delhaye, J.: 1965, in A. Blaauw and M. Schmidt (eds.), *Galactic Structure*, Chicago University Press, Chicago, p. 61.
[2] Stromberg, G.: 1924, *Astrophys. J.* **59**, 228; 1925, *Astrophys. J.* **61**, 363.
[3] Parenago, P. P.: 1950, *Astron. Zh.* **27**, 150, German translation in *Abhandlungen aus der Sowjetischen Astronomie*, Series II, Verlag Kultur und Fortschritt, D.D.R., Berlin 1951, p. 57.
[4] Blaauw, A.: 1965, in A. Blaauw and M. Schmidt (eds.), *Galactic Structure*, Chicago University Press, Chicago, p. 435.
[5] Lindblad, B.: 1927, *Arkiv Mathematik, Astron. Fysik*, **19A**, No. 21, Stockholm.
[6] Lindblad, B.: 1926, *Vierteljahresschrift Astron. Ges.* **61**, 265.
[7] Oort, J. H.: 1926, *Publ. Kapteyn Astron. Lab.* No. 40, Groningen.
[8] Hoffleit, D.: 1964, *Catalogue of Bright Stars*, Yale University Observatory, New Haven.

TABLE VI.2[a]

Space velocities of main sequence G-type stars with known UBV Color indexes

HD	U	V	W	HD	U	V	W
225239	+ 100.9	− 41.2	− 9.8	88725	− 53.3	− 18.8	− 22.3
1581	+ 62.2	− 4.0	− 38.8	90508	− 18.5	− 79.8	+ 18.9
3628	+ 146.9	− 50.5	+ 45.4	95128	+ 24.2	− 2.2	+ 2.1
4614	+ 30.0	− 9.7	− 15.6	98231	+ 0.2	− 22.5	− 19.1
6582	+ 40.7	− 154.9	− 33.9	102158	+ 87.3	− 89.3	+ 12.2
10145	+ 93.0	− 50.3	− 15.6	106116	+ 108.0	− 5.8	+ 25.5
10307	+ 36.6	− 28.1	− 0.2	109358	+ 33.3	− 3.8	+ 1.6
13043	− 19.6	− 60.9	+ 23.2	110897	+ 37.2	+ 8.1	+ 76.4
BD − 1°306	+ 165.6	− 164.5	+ 50.2	114710	+ 45.3	+ 10.5	+ 8.2
13974	+ 36.6	− 50.0	+ 11.7	115043	− 14.5	+ 1.8	− 7.3
14412	+ 10.0	+ 23.3	− 9.0	115383	+ 29.9	+ 2.2	− 18.8
18757	+ 76.4	− 92.6	− 31.5	117176	− 14.2	− 57.1	− 5.6
19467	− 17.2	− 30.0	− 20.2	117635	+ 99.1	− 13.3	− 20.9
19373	+ 81.3	− 21.7	+ 25.6	120593	+ 13.7	− 20.5	+ 0.6
20619	+ 23.6	− 30.0	− 4.0	120690	+ 31.9	− 51.0	− 7.8
20630	+ 22.8	− 4.9	− 4.1	126053	− 21.3	− 14.6	− 37.3
23050	+ 56.2	− 41.2	− 0.7	127356	− 55.3	− 24.1	− 25.1
25680	+ 26.5	− 10.9	− 7.7	130145	− 63.4	− 34.1	− 7.6
29587	+ 144.0	− 73.6	+ 18.1	131156	− 6.0	+ 1.2	+ 1.1
30455	+ 52.1	− 102.3	− 33.6	134331	+ 2.9	− 1.6	− 11.4
30649	+ 77.7	− 135.3	− 14.1	135101	+ 80.8	− 32.1	+ 11.4
34411	+ 77.0	− 45.1	+ 5.4	136352	+ 131.9	− 61.7	+ 44.2
38858	+ 9.9	− 36.5	− 13.7	137107	− 15.9	− 5.2	− 12.7
39587	− 13.6	+ 3.2	− 9.2	141004	+ 48.8	− 23.0	− 40.5
42618	− 63.0	− 13.7	+ 11.5	143761	− 71.4	− 52.7	+ 24.3
53705	+ 51.5	− 72.6	− 20.6	146233	− 31.6	− 19.3	− 31.8
55575	+ 81.0	− 5.2	+ 32.3	147584	− 13.4	+ 4.8	− 6.7
64096	− 25.4	+ 0.3	− 18.7	149105	− 36.2	− 61.5	− 15.0
66171	+ 146.2	− 98.4	− 23.0	152792	− 58.5	− 2.3	− 13.1
68017	+ 46.4	− 61.4	− 41.6	156802	+ 63.3	− 75.2	− 56.8
60298	− 131.4	− 3.8	− 46.4	156968	− 37.4	− 61.6	− 30.0
71881	+ 32.1	− 50.1	0.0	157089	+ 158.9	− 46.4	− 24.4
72905	− 10.2	+ 2.9	− 10.2	157214	− 26.0	− 81.1	− 63.7
73393	+ 82.0	− 62.3	− 15.5	160269	− 38.9	− 0.5	− 21.7
73668	− 54.6	− 35.2	− 7.4	165401	+ 83.4	− 89.0	− 39.4
78558	+ 53.0	− 62.5	− 39.1	177758	− 38.2	− 92.9	− 1.6
81809	+ 38.0	− 47.1	+ 4.1	179484	− 46.4	− 1.3	+ 46.8
86728	+ 56.8	− 46.0	+ 19.6	179558	− 43.5	+ 14.9	− 38.6
87998	− 16.4	− 50.4	− 58.2	179958	+ 58.7	− 42.1	+ 28.1
88371	+ 122.8	− 21.3	+ 17.3	184499	+ 63.1	− 158.6	+ 54.8

[a] The first column contains the number of the star in the Henry Draper Catalogue (the ninth star's designation is that in the Bonner Durchmusterungen). The velocity components are given in km/s. The data are from Eggen's Catalogue of Space-Velocities (Royal Observatory Bulletin No. 51).

Table VI.2 (continued)

HD	U	V	W	HD	U	V	W
184700	− 42.2	− 97.1	− 22.8	215812	− 51.9	− 45.7	+ 13.3
186408	− 13.1	− 28.8	− 1.3	217014	+ 13.5	− 27.6	+ 15.0
187923	− 57.0	− 64.5	+ 26.3				
189340	− 54.1	− 40.1	+ 1.4	221914	+ 88.4	− 38.6	+ 10.4
190406	− 39.7	− 19.7	+ 8.7	222794	+ 60.0	− 95.3	+ 73.0
				223238	+ 69.3	− 50.9	− 12.2
193664	+ 35.0	− 5.7	− 18.6	224383	+ 65.3	− 74.3	+ 1.2
201889	+ 141.8	− 78.9	− 41.9	224930	+ 8.5	− 73.0	− 32.0
211476	+ 94.7	− 35.7	− 31.9				

TABLE VI.3[a]

Stars from the Yale *Catalogue of Bright Stars*
with radial velocities larger than 80 km/s in absolute value

BS	l	b	ϱ	BS	l	b	ϱ
316	125°.1	− 5°.9	− 96	4800	126°.5	+ 57°.5	− 91
731	154 .0	− 39 .1	− 116	5535	352 .1	+ 48 .8	+ 83
1963	203 .5	− 14 .6	+ 88	5664	319 .7	− 2 .9	+ 88
1996	237 .3	− 27 .1	+ 110	5730	313 .8	− 14 .0	+ 96
2018	177 .9	+ 2 .7	+ 103	6128	7 .4	+ 27 .2	+ 100
2028	176 .5	+ 3 .8	+ 100	6272	344 .1	+ 1 .5	− 138
2065	217 .3	− 17 .8	+ 87	6282	350 .2	+ 6 .1	− 92
2140	232 .3	− 21 .6	+ 183	6296	9 .0	+ 19 .3	− 102
2149	238 .4	− 23 .4	+ 94	6364	43 .0	+ 32 .5	− 96
2153	171 .7	+ 10 .0	− 87	7120	12 .9	− 10 .9	− 110
2574	224 .0	− 4 .9	+ 97	7127	335 .9	− 24 .2	+ 180
2721	170 .3	+ 23 .5	+ 85	7405	59 .0	+ 3 .4	− 86
2878	255 .7	− 11 .9	+ 88	7477	81 .9	+ 13 .2	− 85
3906	235 .0	+ 40 .6	+ 97	7523	74 .8	+ 8 .2	− 97
3999	284 .9	− 4 .5	+ 289	7957	97 .8	+ 11 .6	− 87

[a] The first column contains the number of the star in the Yale Catalogue. The new galactic coordinates stand in the second and third column. The radial velocity, ϱ, in km/s, is given in the fourth column.

THE THEORY OF GALACTIC ROTATION

1. In his paper "Observational evidence confirming Lindblad's hypothesis of a rotation of the galactic system", published in 1927 [1], and which marked a milestone in the evolution of our ideas on the dynamics of the stellar system, Prof. J. H. Oort wrote:

Lindblad has recently put forward an extremely suggestive hypothesis, giving a beautiful explanation of the general character of the systematic motions of the stars of high velocity. He supposes that the greater galactic system ... may be divided up into sub-systems each of which is symmetrical around the axis of symmetry of the greater system and each of which is approximately in a state of dynamical equilibrium. The sub-systems rotate around their common axis, but each one has a different speed of rotation.

In a footnote Prof. Oort adds:

Of course the rotation considered is not generally one of constant angular velocity throughout the sub-system. In the following comparisons between the speeds of rotation these speeds are taken for stars at the same distance from the axis.

He then goes on:

One of these sub-systems is defined by the globular clusters for instance; this one has a very low speed of rotation. The stars of low velocity observed in our neighbourhood form part of another sub-system. As the rotational velocity of the slow moving stars is about 300 km/s and the random average velocity only 30 km/s, these stars can be considered as moving very nearly in circular orbits around the centre ...

The following paper is an attempt to verify in a direct way the fundamental hypothesis underlying Lindblad's theory, namely that of the rotation of the galactic system around a point near the centre of the system of globular clusters.

In the present exercise we shall make, in an abridged and simplified form, the same attempt. First, in the following sections, we shall discuss some general points and derive the fundamental formula describing the effect of galactic rotation on radial velocities. In view of the rather large amount of computational work involved, the present exercise will be divided up into two parts. In the first one we shall follow the line of approach adopted by Oort in his paper referred to above. The second part deals with a method of analysis devised and applied by G. L. Camm.

2. Apart from a special case, namely that of the Hyades moving cluster, we have so far investigated only the motions of the stars in the vicinity of the Sun. The theory of galactic rotation represents, on the contrary, an attempt to explain the kinematics of the whole stellar system. Nevertheless some of the arguments previously used can serve as a good starting point for an account of the theory of galactic rotation.

Considering, in Exercise IV, the motions of a group of stars, we have introduced the notion of the centroid as of the point moving with a velocity $\mathbf{V}_c$ defined by

$$\mathbf{V}_c = \frac{1}{n} \sum_i \mathbf{V}_{\mathrm{sp},i} \tag{VII.1}$$

where n is the number of stars considered, and where $\mathbf{V}_{\mathrm{sp},i}$ is the space velocity of the i-th star of the group. In other words the velocity of the centroid characterises what may be called the mean motion of a group of stars contained in an elementary volume of space. The residual velocities, $\mathbf{V}_{\mathrm{res},i}$ have been defined by:

$$\mathbf{V}_{\mathrm{res},i} = \mathbf{V}_{\mathrm{sp},i} - \mathbf{V}_c, \quad i = 1, 2, ..., n. \tag{VII.2}$$

The space velocities used so far were, in fact, velocities referred to a frame having the Sun as its origin and, therefore, moving with the Sun. Now we will adopt a frame at rest, having its origin in the center of the galactic system. Let us retain, for the velocities referred to this new frame, the same notations, and rewrite the relation which corresponds to (VII.2) as follows:

$$\mathbf{V}_{\mathrm{sp},i} = \mathbf{V}_c + \mathbf{V}_{\mathrm{res},i}. \tag{VII.3}$$

At this point the following important remark must be made. In any elementary volume the residual velocity will vary in an irregular, random way when one passes from one star to another, as it is a quantity which describes the motions of the stars as seen on a microscopic level. On the contrary the velocity of the centroid has, per definition, the same value for all the stars of the given group and contained in the volume element considered. This quantity describes – one may say – the motion as seen on a macroscopic level. In other words by Equation (VII.3) we have, in fact, decomposed the velocity of any star into two essentially different components.

In the theory of galactic rotation we are concerned only with the first of these two components. Using mathematical terms, we may say that this theory specifies the field of the vector $\mathbf{V}_c$. Effectively, when it is said that the theory of galactic rotation postulates that:

"1. The mean motion of the stars in any small volume centered in the galactic plane is that of a circular orbit around the galactic center. The plane of the orbit coincides with the galactic plane ..." and that,

"2. The mean motion of stars in a volume whose center S lies at a moderate distance z from the galactic plane is practically the same as that prevailing near the projection of S on the galactic plane";
this means that in the first case the corresponding centroids describe in the galactic plane circular orbits around the galactic center; whereas in the second they describe, in the plane parallel to the galactic plane, circular orbits around the symmetry axis of the Galaxy. The last sentence in 2. states that the velocity depends only on the distance from this axis. The quotation is from Trumpler's and Weaver's monograph.

3. Any attempt to test fully the theory of galactic rotation involves a discussion of the radial velocities as well as of the proper motions of stars (or other galactic objects).

It also presupposes the knowledge of their positions in space. For practical reasons, especially due to the fact that the radial velocities are less exposed to systematic errors than the proper motions, most of the investigations have so far been based on radial velocities only. The values of some parameters must then be taken from other investigations. In the present exercise we also shall use radial velocities.

Let us now derive the formula expressing the effect of galactic rotation on the radial velocities. For the derivation of the complete set of formulae (which set comprises also the expression for the effect on the proper motions) the reader is referred to the problems at the end of this exercise.

First we shall define more precisely the fixed frame referred to above. Its origin coincides with the galactic center. Its z-axis is identical to the galactic symmetry axis, i.e. it is drawn through the center, and is perpendicular to the galactic plane. The positive direction of the z-axis is towards the North Galactic Pole. We shall neglect the (very small) distance of the Sun from the galactic plane, and draw the x-axis from the galactic center through the position of the Sun at a given moment. The y-axis shall be chosen so that the system becomes a right-handed one. We shall provisionally call this system of axes the galactic frame (Figure VII.1).

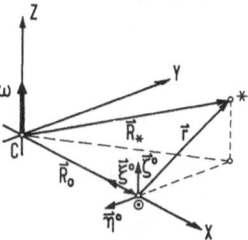

Fig. VII.1.

There is apparently no general agreement concerning the use of right- or left-handed systems in stellar kinematics. In the present exercise we will adopt, for sake of convenience, a right-handed system. Our reasons are the following ones:

1. All currently used mathematics, insofar as it depends on the orientation of a coordinate system, is derived for right-handed systems.

2. The main two systems of spherical coordinates (the equatorial and the galactic one), as well as the systems of space coordinates derived by adding the distance from the origin as a third coordinate, are right-handed ones.

It is true that in a right-handed system the angular velocity of rotation will have a negative sign. However this is, in our opinion, a negligible drawback when weighted against the confusion which, in the beginning, one encounters, when, on the one side directions are defined with respect to right-handed systems, whereas the velocity components are given relative to left-handed axes. In any case after gaining some experience the reader will have no difficulty in passing from formulae derived in one of the two systems to those valid in the other one.

Consider now the stars (or other objects) belonging to a definite class. Let ∗ be such a star and $\mathbf{R}_*$ its radius vector with respect to the galactic center C (Figure VII.1). Then according to the theory of galactic rotation and using a formula well known from elementary kinematics we have, for the velocity of the centroid coinciding with the instantaneous position of the star,

$$\mathbf{V}_{c*} = \boldsymbol{\omega} \times \mathbf{R}_* \tag{VII.4}$$

$\boldsymbol{\omega}$ is the angular velocity of galactic rotation at the star distance from the z-axis.

Let $\mathbf{V}_{res}$ be the star's residual velocity. Then, for the space velocity of the star, with respect to the galactic frame, we will have:

$$\mathbf{V}_* = \mathbf{V}_{c*} + \mathbf{V}_{res} = \mathbf{V}_{res} + \boldsymbol{\omega} \times \mathbf{R}_* . \tag{VII.5}$$

Similarly we shall have for the velocity of the centroid coinciding with the position of the Sun

$$\mathbf{V}_{c\odot} = \boldsymbol{\omega}_0 \times \mathbf{R}_0 \tag{VII.6}$$

and for the Sun's velocity relative to the galactic frame

$$\mathbf{V}_\odot = \mathbf{V}_{c\odot} + \mathbf{S} = \mathbf{S} + \boldsymbol{\omega}_0 \times \mathbf{R}_0 \tag{VII.7}$$

where $\mathbf{S}$ is the Sun's velocity with respect to the centroid at $\odot$.

The radial velocity is, by definition, the component, along the sight line, of the velocity relative to the observer. The radial velocities listed in catalogues and other publications refer always to the Sun. In order to find the radial velocity of the star let us first determine its velocity with respect to the Sun. From (VII.5) and (VII.7) we have, for the relative velocity,

$$\mathbf{V} = \mathbf{V}_* - \mathbf{V}_\odot = \mathbf{V}_{res} - \mathbf{S} + \boldsymbol{\omega} \times \mathbf{R}_* - \boldsymbol{\omega}_0 \times \mathbf{R}_0 . \tag{VII.8}$$

Let, further, $\mathbf{r}$ be the radius vector of the star with respect to the Sun. Obviously

$$\mathbf{r} = \mathbf{R}_* - \mathbf{R}_0 , \qquad \mathbf{R}_* = \mathbf{r} + \mathbf{R}_0 . \tag{VII.9}$$

Substituting the value of $\mathbf{R}_*$ from the second equation in (VII.8) we get for the relative velocity

$$\mathbf{V} = \mathbf{V}_{res} - \mathbf{S} + \boldsymbol{\omega} \times \mathbf{r} + (\boldsymbol{\omega} - \boldsymbol{\omega}_0) \times \mathbf{R}_0 . \tag{VII.10}$$

Take now a new system with the Sun as the origin. Let $\boldsymbol{\xi}^0$, $\boldsymbol{\eta}^0$, $\boldsymbol{\zeta}^0$, be the unit vectors of its axes directed respectively towards the galactic center ($l=0°$, $b=0°$), the point $l=90°$, $b=0°$, and the North Galactic Pole (see Figure VII.1). Then for the angular velocities we will have

$$\boldsymbol{\omega} = \omega \boldsymbol{\xi}^0 , \qquad \boldsymbol{\omega}_0 = \omega_0 \boldsymbol{\xi}^0 \tag{VII.11}$$

where ω and ω_0 are the absolute values of the angular velocities at the star's and at the Sun's distance from the galactic axis respectively. Notice, however, that we shall have

$$\mathbf{R}_0 = - R_0 \boldsymbol{\xi}^0 , \tag{VII.12}$$

R_0 being the Sun's distance from the galactic center.

The coordinate system just introduced is, in fact, identical to the new system of galactic coordinates, the distance r from the Sun being added as a third coordinate.

In this system the direction towards the star, or the sight line is defined by the unit vector

$$\mathbf{r}^0 = \mathbf{r}/r = \cos b \cos l \, \boldsymbol{\xi}^0 + \cos b \sin l \, \boldsymbol{\eta}^0 + \sin b \, \boldsymbol{\xi}^0 \qquad \text{(VII.13)}$$

where l, b, are the galactic coordinates of the star.

In order to find the radial velocity of the star we have to take the scalar product of the relative velocity $\mathbf{V}$ with $\mathbf{r}^0$. Making use of the well-known relations

$$\mathbf{r}^0 \cdot (\boldsymbol{\omega} \times \mathbf{r}) = \boldsymbol{\omega} \cdot (\mathbf{r} \times \mathbf{r}^0) = 0, \qquad \boldsymbol{\zeta}^0 \times \boldsymbol{\xi}^0 = \boldsymbol{\eta}^0 \qquad \text{(VII.14)}$$

we find, from (VII.10) and (VII.13)

$$\mathbf{V} \cdot \mathbf{r}^0 = \varrho = \mathbf{r}^0 \cdot \mathbf{V}_{\text{res}} - \mathbf{r}^0 \cdot \mathbf{S} - (\omega - \omega_0) \, R_0 \cos b \sin l \qquad \text{(VII.15)}$$

where, as usual, ϱ stands for the radial velocity.

Denote now by u_0, v_0, w_0, the components of the Sun's velocity $\mathbf{S}$ along the axes ξ, η, ζ, i.e. put

$$\mathbf{S} = u_0 \boldsymbol{\xi}^0 + v_0 \boldsymbol{\eta}^0 + w_0 \boldsymbol{\zeta}^0 . \qquad \text{(VII.16)}$$

Then

$$\mathbf{r}^0 \cdot \mathbf{S} = u_0 \cos b \cos l + v_0 \cos b \sin l + w_0 \sin b \qquad \text{(VII.17)}$$

and Equation (VII.15) becomes

$$\varrho - \delta\varrho + u_0 \cos b \cos l + v_0 \cos b \sin l + w_0 \sin b$$
$$= - (\omega - \omega_0) \, R_0 \cos b \sin l, \qquad \text{(VII.18)}$$

where $\delta\varrho$ stands for the scalar product $\mathbf{r}^0 \cdot \mathbf{V}_{\text{res}}$.

Before going further remember that, owing to the character of the residual velocities, $\delta\varrho$ will vary in a random way as one passes from one star to another. In other words, the regularity, or the systematic effect, due to galactic rotation, which we expect to find in the radial velocities, will, in fact be given by the equation

$$\varrho' = \varrho + u_0 \cos b \cos l + v_0 \cos b \sin l + w_0 \sin b$$
$$= - (\omega - \omega_0) \, R_0 \cos b \sin l \qquad \text{(VIII.19)}$$

the omitted term $\delta\varrho$ being responsible only for a certain scatter of the radial velocities actually observed around the values predicted by Equation (VII.19).

Equation (VII.19) is the fundamental formula referred to at the end of Section 1. Notice that, according to its derivation, this formula describes the effect of the difference in rotational motions or, as it is sometimes said, the effect of differential motion (or rotation) on the radial velocity.

4. Before discussing the application of the fundamental formula we must make some additional remarks.

As already said, our analysis will be based on the radial velocities of stars. Therefore for each star we intend to use we must know the radial velocity ϱ. For all such stars the galactic coordinates will also be known.

As to the components of the solar velocity, i.e. the quantities u_0, v_0, and w_0, their values could, in principle, be derived from the same analysis. However in view of the volume of computational work involved, we shall refrain from such a determination and use values derived by D. N. W. Stibbs in an important paper [2].

The expression on the right side of Equation (VII.19) will be discussed later. Notice however that it implicitly contains the distance r of the star from the Sun. In fact, according to the fundamental assumption of the theory of galactic rotation, the angular velocity of rotation is a function of the distance from the symmetry axis of the Galaxy. Therefore, instead of ω and ω_0 we, in fact, shall write $\omega(R)$ and $\omega(R_0)$, respectively, where R denotes the star's distance from the symmetry axis. Now from Figure VII.2, one easily finds, applying the cosine theorem of plane trigonometry,

$$R^2 = R_0^2 + (r \cdot \cos b)^2 - 2 \cdot R_0 \cdot r \cdot \cos b \cdot \cos l. \tag{VII.20}$$

Therefore for all stars to be used in our analysis the distances from the Sun must also be known.

Let us now discuss the choice of stars (or other objects) which we can use to test the theory.

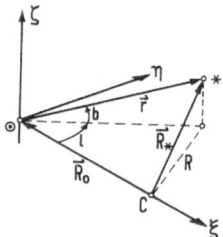

Fig. VII.2.

If we take the degree of flattening of a sub-system as an indication of its speed of rotation, then, obviously, we are led to choose objects belonging to the very flat sub-systems. Such a choice offers, moreover, the advantage that, as known from investigations of the motions of stars in the solar neighborhood, the scatter of the residual velocities is in general quite small in very flattened systems. In other words for such objects the term $\delta\varrho$ will have a small value.

The mean velocity is a quantity which varies smoothly as one passes from one point to another, neighboring one. Or, using mathematical language, the velocity of the centroid is a continuous function of the position in space. It is therefore obvious that the differential effects of the galactic rotation, as described by Equation (VII.19) will, in general, increase with the distance from the Sun of the object considered. We also can say that the systematic effect due to galactic rotation will emerge from the fluctuations due to the random character of the residual velocities only if sufficient distant objects are used. As in this case the trigonometric parallaxes can be ruled out as a source of data on distances, our work will obviously depend on some method of photometric distance determination. In this context it is important to realise that, due to strong and irregular interstellar absorption, the photometric determination of distances of objects close to the galactic plane encounters many difficulties.

In principle for such investigations one can use, and one has used, classical Cepheids, B stars and galactic clusters. Most of the earlier studies have, in fact, been done with Cepheids. In an important paper by R. P. Kraft and M. Schmidt [3] the advantages of using Cepheids are explained as follows:

First, if our adopted Period-Color and Period-Luminosity relations are systematically correct, cepheid distances can be determined with a precision equivalent to that attainable for galactic clusters from photoelectric techniques. Indeed, for a galactic cluster and a cepheid of the same age the cepheid can be used with greater confidence at large distance because it is brighter than any star in the cluster and therefore brighter than the stars used for the main sequence fitting procedure, and is free, as well, from cluster crowding. Second, the cepheid velocities can be determined precisely because the lines are sharp in comparison with most B-stars. Third, cepheids at large distances can easily be picked up by their variability. Until narrow-filter (i.e. narrow-band) photometry for B-stars in the manner of Strömgren and its associates is available, from which it is hoped that a fine scale of absolute magnitudes and spectral types can be set up, we believe that cepheids provide the best available tools for the study of galactic rotation and structure.

In the present exercise we shall use Cepheids. The data we need for our analysis are collected in Table VII.1 at the end of this Exercise. This table represents an excerpt from a longer list given in the paper by Kraft and Schmidt cited. From all Cepheids listed by these authors we have retained only those which in the *General Catalogue of Variable Stars* [4] are classified as classical Cepheids. See Preface to Vol. II of the General Catalogue as well as list C, on pp. 11 and 12, where all such stars are indicated by a δ. The radial velocities have been taken from the paper by Stibbs already cited.

5. In our analysis of the radial velocities we shall use two different methods. However in either case, and for each star, we must beforehand compute the sum standing on the left side of Equation (VII.19). As easily seen from the derivation of this formula, the sum of the trigonometric terms

$$u_0 \cos b \cos l + v_0 \cos b \sin l + w_0 \sin b \qquad \text{(VII.21)}$$

represents the correction to be applied to the radial velocities relative to the Sun, in order to get the radial velocities relative to the centroid coinciding with the position of the Sun.

The values of the trigonometric function needed are given in Table VII.1. For the components of the solar motion we shall adopt the following values:

$$u_0 = +9 \qquad v_0 = +12 \qquad w_0 = +7 \quad \text{(km/s)}$$

which are the values found by Stibbs, rounded off to the nearest whole number.

The computation of these corrections does not present any difficulty, yet it is a rather long undertaking. Quite acceptable approximate values can easily and quickly be found by the following graphic method.

As seen from Table VII.1, the galactic latitudes of practically all the Cepheids are very low. This means that with fair approximation the corrective term will simply be

$$u_0 \cos l + v_0 \sin l \qquad \text{(VII.22)}$$

or, what amounts to the same, that only the component of the solar motion in the galactic plane is of importance. Now decompose the Sun's velocity $\mathbf{S}$ into two components, one $\mathbf{S}_g$ in the galactic plane, the other one, $\mathbf{S}_n$, perpendicular to it, i.e. put

$$\mathbf{S} = \mathbf{S}_g + \mathbf{S}_n. \tag{VII.23}$$

Then, neglecting the very low galactic latitude of the Cepheids used, one can assume that all the unit vectors $\mathbf{r}^0$ pointing from the Sun to any of them are, too, in the galactic plane. According to the definition of the scalar product of two vectors, and taking into account that $\mathbf{r}^0$ is by definition a unit vector, the correction to be applied is, with the approximation adopted, equal to $\mathbf{r}^0 \cdot \mathbf{S}_g$, or, again, equal to the projection of $\mathbf{S}_g$ onto the straight line defined by $\mathbf{r}^0$.

In order to determine this projection compute the absolute value as well as the direction of $\mathbf{S}_g$. They are given by the obvious equations

$$|\mathbf{S}_g| = [u_0^2 + v_0^2]^{1/2} \text{ km/s} ; \qquad \text{tg} \, l_0 = v_0/u_0. \tag{VII.24}$$

Now draw a circle having a radius equal to S_g, taking, for convenience, 1 cm for 1 km/s (Figure VII.3), and mark along the circumference the galactic longitudes. Then for a star at galactic longitude l, the correction sought for will be given by $\overline{OA}$ or by $\overline{OB}$ which last segment can easily be read out from our simple nomogram.

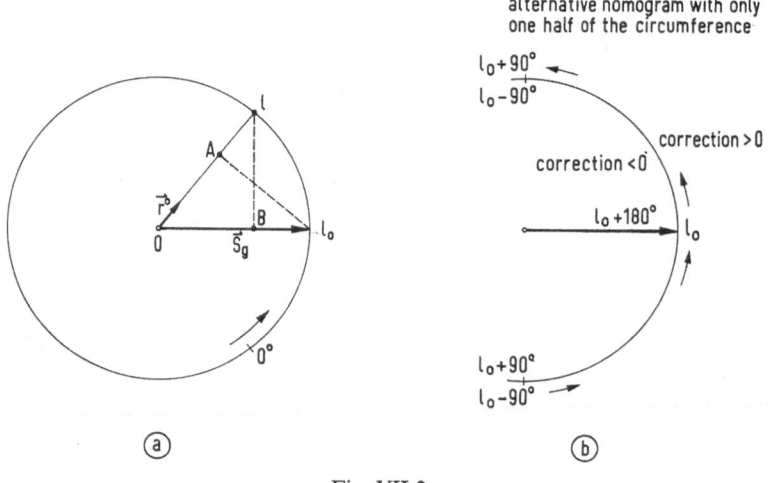

Fig. VII.3.

Notice that for galactic longitudes on the left half of the circle the corrections are negative, as the angle between r^0 and S_g comes in the interval 90° to 270°.

The corrections found in this way correspond to the approximate formula (VII.22), for it has been assumed that

$$\mathbf{r}^0 = \cos l \cdot \mathbf{\xi}^0 + \sin l \cdot \mathbf{\eta}^0. \tag{VII.25}$$

Compute, or determine from the nomogram, the corrections to be applied for each star you will use in your analysis. If the nomogram has been used, the differences between the values of ϱ' given in Table VII.1 (pp. 104–106) and those found with the aid of the nomogram shall not exceed about 1 km/s.

6. Let us now see how the data collected in Table VII.1 can be used to disclose the effect of galactic rotation on the radial velocities.

In the first part of this exercise we shall follow the line of approach adopted by Oort in his paper published in 1927.

Consider the expression on the right side of Equation (VII.19) which we will write as follows:

$$- \left[\omega(R) - \omega(R_0) \right] R_0 \cos b \sin l \tag{VII.26}$$

where $\omega(R)$ and $\omega(R_0)$ respectively are the angular velocity at the star's and at the Sun's distance from the symmetry (or rotational) axis, whereas l and b are the galactic coordinates of the star. Remember that, at the time Oort undertook his analysis, nothing was known about the dependence of the angular velocity on the distance. Also, the distance of the Sun from the galactic center, R_0, was known only to the order of magnitude. It is therefore obvious that under such circumstances the exact expression (VII.19) could not be used in a first attempt to verify the theory. If, however we consider only stars at distances r from the Sun so small that all terms of second and higher order in r/R can be neglected, then, expanding $\omega(R)$ into a Taylor series around $R = R_0$ we get

$$\omega(R) = \omega(R_0) + (R - R_0)\,\omega'(R_0), \quad \text{or}$$
$$\omega(R) - \omega(R_0) = (R - R_0)\,\omega'(R_0). \tag{VII.27}$$

This is the first of the two approximations we shall make. The second approximation is the following one. From (VII.20) we have

$$R = R_0 \left[1 + \left(\frac{r \cos b}{R_0} \right)^2 - 2\,\frac{r}{R_0} \cos b \cos l \right]^{1/2}. \tag{VII.28}$$

Expand the expression under the square root according to the binomial theorem neglecting again the terms of second and higher order in r/R. Then

$$R = R_0 - r \cos b \cos l, \quad \text{or} \quad R - R_0 = - r \cos b \cos l. \tag{VII.29}$$

Substituting (VII.27) and (VII.29) in (VII.19) we get

$$\varrho' = \varrho + u_0 \cos b \cos l + v_0 \cos b \sin l + w_0 \sin b$$
$$= \omega'(R_0)\,R_0 r \sin l \cos l \cos^2 b. \tag{VII.30}$$

Making use of the goniometric relation

$$\sin 2l = 2 \sin l \cos l \quad \text{or} \quad \sin l \cos l = \tfrac{1}{2} \sin 2l, \tag{VII.31}$$

Equation (VII.30) can be put into the form

$$\varrho' = A r \sin 2l \cos^2 b \tag{VII.32}$$

where

$$A = \tfrac{1}{2}\omega'(R_0) \cdot R_0.$$ (VII.33)

This is Oort's first equation, A is Oort's first constant. For the derivation of the second equation as well as the definition of the second constant see Problem VII.3.

Consider now a group of stars all at the same distance r from the Sun, but distributed over all longitudes. Suppose, for sake of simplicity, that their galactic latitudes are equal to zero. Then $\cos^2 b$ is equal to unity and Equation (VII.32) reduces to

$$\varrho' = Ar \sin 2l.$$ (VII.34)

Neglecting the scatter due to the residual velocities as well as to the observational errors, their radial velocities corrected for solar motion, ϱ', shall be a periodic function of the galactic longitude l, with a period equal to 180°. The radial velocities shall vanish for longitudes equal to 0°, 90°, 180°, and 270°, and attain extreme values at the longitudes 45°, 135°, 225°, and 315°. Within the limits of validity of the approximations made, the amplitude will increase linearly with the distance. This is the so-called double wave effect of galactic rotation on the radial velocities.

Now it is obvious that we never shall find a sufficient number of stars having practically the same distance from the Sun. Therefore when testing the theory we shall be obliged to use stars whose distances are spread over a more or less wide interval. Moreover the galactic latitudes of the stars will be different from zero. In this case it is the quantity $\varrho'/r \cdot \cos^2 b$ which will be a periodic function of the galactic longitude l, as, in fact, Equation (VII.32) can be written as

$$\varrho'/r \cdot \cos^2 b = A \sin 2l.$$ (VII.35)

However, for our present needs the galactic latitudes can safely be neglected and $\cos^2 b$ put equal to unity. In this case we shall, in fact, make use of the relation:

$$\varrho'/r = A \sin 2l.$$ (VII.36)

Note that the extreme values as well as the zeros will be attained at the same galactic longitudes as before, but that now the amplitude will simply be equal to A.

In order to see whether or not such an effect is present, proceed as follows. From Table VII.1 pick out, and make a list of, all the Cepheids at distances from the Sun $1.4 \leqslant r < 2.0$. For distances above 2.0 kpc the approximations adopted break down; below about 1.4 kpc the effect is masked by large fluctuations. The columns in your list shall be arranged as follows:

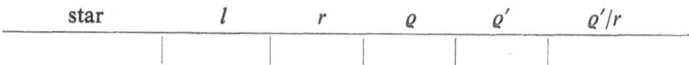

star	l	r	ϱ	ϱ'	ϱ'/r

For each star determine from the nomogram the correction to be applied to the radial velocity ϱ in order to get ϱ', and then compute the corrected radial velocities ϱ'. Finally compute for each star the quantity ϱ'/r (which one may call the corrected radial velocity reduced to unit distance).

Now make a diagram with the values of ϱ'/r on the vertical axis, and the galactic longitudes on the horizontal axis. Figure VII.4 shows schematically the kind of diagram you will get.

As you will see from your graph, the distribution of the points corresponds quite well to the double wave as predicted by the theory. Therefore we can conclude that the effect described by Oort's formula is indeed present.

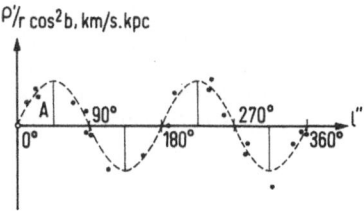

Fig. VII.4.

Our next task is to determine the value of the first constant A. This is usually made by the method of least squares, considering the Equations (VII.32) as the equations of condition. For the moment we shall limit ourselves to the derivation of a rough approximative value of A. In the next section we shall give a method which yields quite acceptable values of A.

Remember that on your graph the points are (assumed to be) distributed along a periodic curve which crosses the horizontal axis at the longitudes 0°, 90°, 180°, and 270°, and which attains extreme values, all equal in absolute value to A, at the longitudes 45°, 135°, 225°, and 315°. See Figure VII.4, where all these longitudes have been marked on the horizontal axis. Make a couple of estimates of the amplitude (A', A''). Draw the corresponding curves ($A' \sin 2l$, $A'' \sin 2l$). Adopt for A the amplitude of this one which gives the best fit.

7. In 1938, G. L. Camm published a paper entitled 'A Study of Galactic Rotation, Based on the Velocities of the Planetary Nebulae' [5]. In this paper Camm developed and applied

An alternative method ... to obtain more, and more direct, information as to how the speed of rotation is related to the distance from the galactic centre.

The galaxy is assumed to rotate in concentric circles, with a speed depending only on the radius of the circle; the most direct procedure is therefore to calculate for each planetary its distance from the galactic centre, and (from its radial velocity) its angular speed of rotation about the centre. After taking average values of radii and angular velocities for nebulae in neighbouring circles, a curve is drawn, giving the relation between the speed of rotation and galactic radius.

From this curve a theoretical value is found for the radial velocity of each planetary; this is seldom in agreement with the observed value, and gives rise to a velocity-dispersion, over and above the mean motion. This velocity dispersion, as measured by the mean square residual velocity, is reduced by making slight corrections to the constants of the problem, of which provisional values had been assumed. Such constants are the distance and the direction of the galactic centre, the components of the solar motion, and a constant K-term. The velocity dispersion, after these corrections have been made, has a smaller value than previously has been obtained for the planetary nebulae.

It is important to notice that in this context the term residual velocity has a some-what different meaning, as the difference between the observed and the mean motion is partly attributed to errors in the assumed values of the constants (or parameters of the problem).

The linear speed of rotation at the distance of the Sun cannot be determined from the radial velocities, but it can be evaluated from the proper motions .... The linear speed at the Sun's distance having been found, a curve is drawn to show the variation of linear speed with galactic radius.

It is assumed that the mean rotation now determined corresponds to circular orbits, described under the gravitational attraction of the whole galaxy; hence it is possible to find the variation of the force with galactic radius. The galaxy is usually supposed to consist of a thin circular disc of matter, with a massive nucleus at the centre. By adjusting the masses of the two parts, it is possible to get a theoretical gravitational field which corresponds very closely to the field of force deduced from the observations.

Stated in very clear terms, Camm's own account of his method scarcely needs any comment. In this second part of the present exercise, we shall make use of Camm's method in a somewhat simplified form.

First of all let us derive the value of Oort's constant A. In doing this we shall follow a procedure applied by Gascoigne and Eggen [6], which effectively represents an intermediary step between Oort's original method and Camm's method. In order to derive the fundamental equation to be used, start again from Equation (VII.19), i.e. from

$$\varrho' = - \left[\omega(R) - \omega(R_0) \right] R_0 \cos b \sin l.$$
(VII.37)

Considering again stars at moderate distances from the Sun, adopt, as before, the first of the two approximations used in deriving Oort's formulae, i.e. Equation (VII.27). However do not make use of the second one. Instead compute by formula (VII.20) for each star the value of $R - R_0$. Taking into account Equation (VII.33) we easily get:

$$\varrho' = - (R - R_0) R_0 \omega'(R_0) \cos b \sin l, \quad \text{or}$$
$$\varrho' = - 2A (R - R_0) \cos b \sin l.$$
(VII.38)

This is our fundamental formula, which, more conveniently can be written as follows:

$$\varrho'/\cos b \cdot \sin l = - 2A (R - R_0).$$
(VII.39)

According to this formula there is a linear relation between the quantity on the left side and the difference $R - R_0$. In other words if, for a group of stars, we make a dia-gram, with the values of $\varrho'/\cos b \cdot \sin l$ on the vertical axis and the values of $R - R_0$ on the horizontal axis, then the points shall be distributed along a straight line drawn through the origin, the slope of which is equal to $-2A$. See Figure VII.5.

Notice, first, that the existence of such a linear relation, when it has been proved, can be considered as a new (and even sharper) test of the theory of galactic rotation. Notice, further, that in order to make use of the method just explained, we must know the value of the Sun's distance from the galactic center, R_0, as well as the direction towards the center, which is implicitly present in the values of the galactic coordinates used. Moreover, when computing the corrected radial velocities, ϱ', we have used

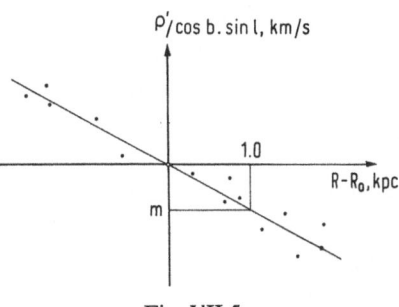

Fig. VII.5.

definite values for the elements of the solar motion (u_0, v_0, w_0). Therefore errors which may be present in the assumed values of any of all these quantities will affect our determination of A. We shall not discuss their influence.

Another very important point is the following one. Due to the presence of the factor $\sin l$ in the denominator on the left side of Equation (VII.39), possible errors in ϱ' will greatly be magnified if $\sin l$ is small. In order to avoid their influence Equation (VII.39) shall, according to Camm, only be applied to stars for which $\sin l$ exceeds $\sin 15°$ in absolute value. Or, what amounts to the same, from our analysis we shall omit stars with galactic longitudes $345° \leqslant l \leqslant 15°$ and $165° \leqslant l \leqslant 195°$.

Among the stars in your preceding list, choose those which fulfil this condition. Make use of the corrected radial velocities ϱ' already derived. By formula (VII.20) compute for each star $R - R_0$ assuming $R_0 = 10$ kpc. Neglect the (in general very small) galactic latitudes, i.e. put $\cos b = 1$. This means that instead of Equation (VII.39) the following one will be used:

$$\varrho'/\sin l = - 2A (R - R_0). \tag{VII.40}$$

Make a list arranged as indicated below. Compute for all the stars the values of

star	l	$\sin l$	ϱ'	$R - R_0$	$\varrho'/\sin l$

$\varrho'/\sin l$, and make the $\varrho'/\sin l$ versus $R - R_0$ diagram. Appreciate to what extent the distribution of the points obeys a linear relation of the form (VII.40). Draw through the origin a straight line fitting at best all the points. Determine its slope m and derive A:

$$A = - m/2, \quad \text{km/s} \cdot \text{kpc}.$$

The value of A adopted at present is 15 km/s·kpc.

8. The application of Camm's method proper shall now present no difficulty. The fundamental equation will now be Equation (VII.19) which more conveniently can be written as follows:

$$- \varrho'/R_0 \cdot \cos b \cdot \sin l = \omega (R) - \omega (R_0). \tag{VII.41}$$

Notice that this equation is exact in the sense that it involves no approximations.

For our present purpose we safely can assume that $\cos b = 1$. Moreover we shall follow the general use and take the angular velocity with a positive sign although it corresponds to clockwise rotation. Therefore our fundamental equation will be

$$\varrho'/R_0 \cdot \sin l = \omega(R) - \omega(R_0). \tag{VII.42}$$

Determine for all the remaining stars in Table VII.1 the corrected radial velocities and compute for each the quantity standing on the left side of Equation (VII.42). Compute for all stars $R - R_0$ as in the preceding section. Make the $\varrho'/R_0 \cdot \sin l$ versus $R - R_0$ (or versus R) diagram. Figure VII.6 shows schematically the kind of diagram you will get.

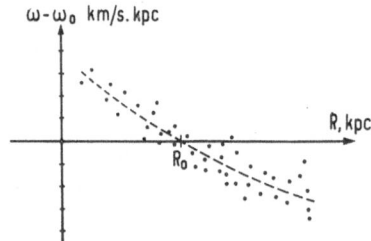

Fig. VII.6.

By Equation (VII.42) $\varrho'/R_0 \cdot \sin l$ is equal to $\omega(R) - \omega(R_0)$. Therefore the points on your diagram define the curve giving $\omega(R) - \omega(R_0)$ in function of the distance from the galactic axis. Notice that this curve must go through the point $(R_0, 0)$. Trace this curve. It gives the so-called angular velocity law. Notice that Oort's first constant can now be derived from the slope of the curve at $R = R_0$. Compare your angular velocity curve to those given in the references. [7] and [8].

Obviously the data as well as the derived quantities shall be arranged in an appropriate form, the choice of which is left to the reader.

If $\omega(R_0) = \omega_0$ is known (see Problem VII.3), the relation between $\omega(R) - \omega(R_0)$ and R can easily be converted into a $\omega(R)$ versus R curve. If, moreover, we interpret the circular motions in the galactic plane as due to an attractive force directed towards the galactic center, then the centripetal acceleration, which is given by $-\omega^2(R) \cdot R$, can be derived from the $\omega(R)$ curve. This problem will briefly be considered in the next Exercise.

It is not difficult to generalise the theory to the case when a general expansion of the galactic system is superimposed on pure rotation, as has been done, e.g. by D. N. W. Stibbs [2]. The terms corresponding to this expansion can easily be derived in the same way as the rotational terms in Problem VII.2.

For a critical review of the different methods which can be used to determine the rotation parameters of the Galaxy the reader is referred to an article by M. Schmidt [9].

PROBLEMS

Problem VII.1. Derivation of the Formulae for the Proper Motion Components in Galactic Longitude and Galactic Latitude

Let α, δ be the equatorial coordinates of a star, and l, b, its galactic coordinates. Let, further, μ_α and μ_δ be the proper motions of the star in right ascension and in declination. Denote by μ_l and by μ_b the respective (annual) variations of l and of b caused by the star's proper motion. These quantities, which are respectively called the proper motion in galactic longitude, and the proper motion in galactic latitude, cannot be derived from observation. Instead they are computed by the formulae

$$\mu_l \cos b = + \mu_\alpha \cos \delta \cos \varphi + \mu_\delta \sin \varphi,$$
$$\mu_b = - \mu_\alpha \cos \delta \sin \varphi + \mu_\delta \cos \varphi. \tag{P.VII.1}$$

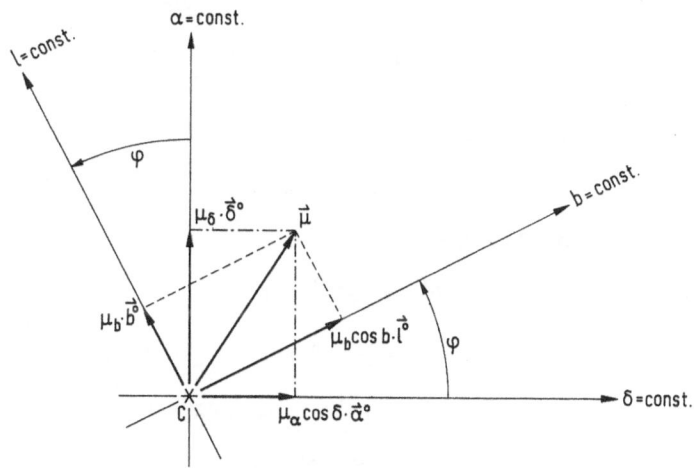

Fig. VII.7.

The meaning of the angle φ is best explained by Figure VII.8. This angle can be computed from the formula

$$\cos \varphi = \frac{\sin \delta_p - \sin \delta \sin b}{\cos \delta \cos b}, \tag{P.VII.2}$$

δ_p is the declination of the North Galactic Pole. Prove the formulae (P.VII.1) and (P.VII.2).

Solution

Consider the total proper motion μ as an elementary vector, as has been done in

Exercise III, Problem III.3. Then (Figure VII.7)

$$\mu = \mu_\alpha \cos\delta \cdot \mathbf{a}^0 + \mu_\delta \cdot \mathbf{\delta}^0 . \qquad\qquad (\text{P.VII.3})$$

Now introduce a unit vector $\mathbf{l}^0$ tangent to the circle of constant galactic latitude at the star's position, and another unit vector $\mathbf{b}^0$ tangent to the circle of constant galactic longitude. Then

$$\mu = \mu_l \cos b \cdot \mathbf{l}^0 + \mu_b \cdot \mathbf{b}^0 . \qquad\qquad (\text{P.VII.4})$$

Notice that, like $\mathbf{a}^0$ and $\mathbf{\delta}^0$, the two unit vectors $\mathbf{l}^0$ and $\mathbf{b}^0$ will also be mutually perpendicular.

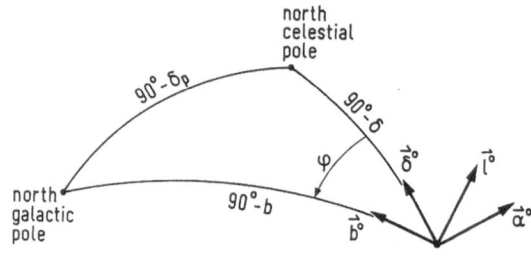

Fig. VII.8.

Now take the scalar product of

$$\mu_l \cos b \cdot \mathbf{l}^0 + \mu_b \cdot \mathbf{b}^0 = \mu_\alpha \cos\delta \cdot \mathbf{a}^0 + \mu_\delta \cdot \mathbf{\delta}^0 \qquad\qquad (\text{P.VII.5})$$

first by $\mathbf{l}^0$ and then by $\mathbf{b}^0$. This gives

$$\mu_l \cos b = \mu_\alpha \cos\delta \cdot (\mathbf{l}^0 \cdot \mathbf{a}^0) + \mu_\delta (\mathbf{l}^0 \cdot \mathbf{\delta}^0) \qquad\qquad (\text{P.VII.6a})$$

and

$$\mu_b = \mu_\alpha \cos\delta\,(\mathbf{b}^0 \cdot \mathbf{a}^0) + \mu_\delta\,(\mathbf{b}^0 \cdot \mathbf{\delta}^0) . \qquad\qquad (\text{P.VII.6b})$$

From Figure VII.7 one easily finds for the scalar products

$$\begin{aligned}
(\mathbf{l}^0 \cdot \mathbf{a}^0) &= +\cos\varphi, & (\mathbf{l}^0 \cdot \mathbf{\delta}^0) &= +\sin\varphi, \\
(\mathbf{b}^0 \cdot \mathbf{a}^0) &= -\sin\varphi, & (\mathbf{b}^0 \cdot \mathbf{\delta}^0) &= +\cos\varphi.
\end{aligned} \qquad (\text{P.VII.7})$$

which proves the Equations (P. VII.1).

The formula for φ follows immediately from the cosine theorem of spherical trigonometry, applied to the triangle North Galactic Pole–North Celestial Pole–the star; see Figure VII.8.

Problem VII.2.   General Formulae for the Effects of Differential Galactic Rotation

Derive the set of formulae describing the effect of galactic rotation on the proper motions in galactic longitude, galactic latitude, and radial velocity. Hint: Make use of relations analogous to Equations (III.7) p. 35, but defined for galactic coordinates.

Solution

In order to derive formulae analogous to (III.7) but valid for galactic coordinates, take a coordinate system having its origin in the Sun, and direct its axes respectively towards the points $l=0°$, $b=0°$; $l=90°$, $b=0°$; and the North Galactic pole. The unit vectors which now define the astronomical trihedral are $\mathbf{l}^0$ and $\mathbf{b}^0$, introduced in the preceding Problem, and $\mathbf{r}^0$ defined in VII.3, Equation (VII.13). The unit vectors of the coordinate axes are $\boldsymbol{\xi}^0$, $\boldsymbol{\eta}^0$, $\boldsymbol{\zeta}^0$.

Let V_l, V_b, and ϱ, respectively be the components of any space velocity along the axes of the trihedral. Denote by u, v, w, its components along the axes of the coordinate system having its origin in the Sun. Then obviously

$$\begin{aligned}
V_l &= - u \sin l \qquad\quad + v \cos l \\
V_b &= - u \sin b \cos l \; - \; v \sin b \sin l \; + \; w \cos b \qquad\qquad \text{(P.VII.8)} \\
\varrho &= + u \cos b \cos l + v \cos b \sin l + w \sin b \, .
\end{aligned}$$

The first two components are related to the galactic proper motions by the evident relations

$$V_l = 4.738 r \mu_l \cos b \, , \qquad V_b = 4.738 r \mu_b \, , \quad \text{km/s} \, . \qquad \text{(P.VII.9)}$$

The third component is simply the radial velocity ϱ.

Notice that the proper motions must be expressed in $''/a$, and the distance r measured in parsecs.

Neglecting the residual velocity of the star, that is, considering only the mean motion, we have from Equation (VII.10) p. 86, for the velocity of the star with respect to the Sun

$$\mathbf{V} = - \mathbf{S} + \boldsymbol{\omega} \times \mathbf{r} + (\boldsymbol{\omega} - \boldsymbol{\omega}_0) \times \mathbf{R}_0 \, . \qquad \text{(P.VII.10)}$$

All we now have to do is to find the components u, v, and w, of the velocity vector $\mathbf{V}$, and then to substitute their values into Equations (P. VII.8).

The components of $-\mathbf{S}$ are $-u_0$, $-v_0$, and $-w_0$. The components of the two vector products can easily be found by making use of the determinant form of the vector product. In fact we have

$$\boldsymbol{\omega} \times \mathbf{r} = \begin{vmatrix} \boldsymbol{\xi}^0 & \boldsymbol{\eta}^0 & \boldsymbol{\zeta}^0 \\ \omega_u & \omega_v & \omega_w \\ r_u & r_v & r_w \end{vmatrix} = \begin{vmatrix} \boldsymbol{\xi}^0 & \boldsymbol{\eta}^0 & \boldsymbol{\zeta}^0 \\ 0 & 0 & \omega \\ r \cos b \cos l & r \cos b \sin l & r \sin b \end{vmatrix} \qquad \text{(P.VII.11)}$$

and from this, for the components along the u-, v-, and w-axes, respectively

$$(\boldsymbol{\omega} \times \mathbf{r})_u = - \omega r \cos b \sin l, \qquad (\boldsymbol{\omega} \times \mathbf{r})_v = + \omega r \cos b \cos l,$$
$$(\boldsymbol{\omega} \times \mathbf{r})_w = 0. \qquad \text{(P.VII.12)}$$

Similarly, for the second vector product,

$$(\boldsymbol{\omega} - \boldsymbol{\omega}_0) \times \mathbf{R}_0 = \begin{vmatrix} \xi^0 & \eta^0 & \zeta^0 \\ 0 & 0 & \omega - \omega_0 \\ -R_0 & 0 & 0 \end{vmatrix}, \qquad \text{(P.VII.13)}$$

one finds:

$$[(\boldsymbol{\omega} - \boldsymbol{\omega}_0) \times \mathbf{R}_0]_u = 0, \qquad [(\boldsymbol{\omega} - \boldsymbol{\omega}_0) \times \mathbf{R}_0]_v = - (\omega - \omega_0) R_0,$$
$$[(\boldsymbol{\omega} - \boldsymbol{\omega}_0) \times \mathbf{R}_0]_w = 0. \qquad \text{(P.VII.14)}$$

Therefore we will have, for the components of the velocity vector $\mathbf{V}$,

$$\begin{aligned} u &= - u_0 + (\boldsymbol{\omega} \times \mathbf{r})_u + [(\boldsymbol{\omega} - \boldsymbol{\omega}_0) \times \mathbf{R}_0]_u = - u_0 - \omega r \cos b \sin l \\ v &= - v_0 + (\boldsymbol{\omega} \times \mathbf{r})_v + [(\boldsymbol{\omega} - \boldsymbol{\omega}_0) \times \mathbf{R}_0]_v = - v_0 + \omega r \cos b \cos l - (\omega - \omega_0) R_0 \\ w &= - w_0 + (\boldsymbol{\omega} \times \mathbf{r})_w + [(\boldsymbol{\omega} - \boldsymbol{\omega}_0) \times \mathbf{R}_0]_w = - w_0. \end{aligned}$$
$$\text{(P.VII.15)}$$

Substituting this in (P. VII.8) and making use of the relations (P. VII.9) we get

$$\begin{aligned} 4.738 r \mu_l \cos b &= + u_0 \sin l \qquad - v_0 \cos l \qquad + \omega r \cos b \\ & \quad - (\omega - \omega_0) R_0 \cos l \\ 4.738 r \mu_b &= + u_0 \sin b \cos l + v_0 \sin b \sin l - w_0 \cos b \\ & \quad + (\omega - \omega_0) R_0 \sin b \sin l \\ \varrho &= - (u_0 \cos b \cos l + v_0 \cos b \sin l + w_0 \sin b) \\ & \quad - (\omega - \omega_0) R_0 \cos b \sin l. \end{aligned}$$
$$\text{(P.VII.16)}$$

Rewrite these equations as follows:

$$\begin{aligned} 4.738 r \mu_l \cos b - u_0 \sin l \qquad + v_0 \cos l \\ = - (\omega - \omega_0) R_0 \cos l + \omega r \cos b \\ 4.738 r \mu_b - u_0 \sin b \cos l - v_0 \sin b \sin l + w_0 \cos b \\ = + (\omega - \omega_0) R_0 \sin b \sin l \end{aligned}$$
$$\text{(P.VII.17)}$$

$$\varrho + u_0 \cos b \cos l + v_0 \cos b \sin l + w_0 \sin b = -(\omega - \omega_0) R_0 \cos b \sin l.$$

The left sides are respectively equal to the components V_l, V_b, and to the radial velocity, each one corrected for solar motion. Now divide the first two equations by $4.738\ r$, leaving the third unchanged. Then:

$$\mu_l \cos b - \frac{u_0}{4.738r} \sin l + \frac{v_0}{4.738r} \cos l$$
$$= - \frac{\omega - \omega_0}{4.738r} R_0 \cos l + \frac{\omega}{4.738} \cos b$$
$$\mu_b - \frac{u_0}{4.738r} \sin b \cos l - \frac{v_0}{4.738r} \sin b \sin l + \frac{w_0}{4.738r} \cos b \qquad \text{(P.VII.18)}$$
$$= + \frac{\omega - \omega_0}{4.738r} R_0 \sin b \sin l$$

$$\varrho + u_0 \cos b \cos l + v_0 \cos b \sin l + w_0 \sin b$$
$$= - (\omega - \omega_0) R_0 \cos b \sin l.$$

The left sides of these equations are respectively equal to the proper motions component in galactic longitude, to the proper motion in galactic latitude, and to the radial velocity, all corrected for solar motion.

Equations (P. VII.18) are the expressions sought for.

Problem VII.3.   Oort's Second Formula, and the Relation between Oort's Constants A and B, and the Angular Velocity ω_0

In a paper entitled 'Investigations Concerning the Rotational Motion of the Galactic System, Together with a New Determination of Secular Parallaxes, Precession and Motion of the Equinox' (*Bull. Astron. Inst. Neth.* **4** (1927), 79), Oort gives the following equation for 'the average transversal velocities, after correction for solar motion...'

$$\overline{\mu'_l} = \frac{A}{4.74} \cos 2\,(l - l_0) \cos b + \frac{B}{4.74} \cos b. \tag{P.VII.19}$$

This is the so-called Oort's second equation. It refers, in fact, to the average proper motion in galactic longitude (corrected for solar motion), of a group of stars at a mean distance r, and situated in a limited area of the sky, centered at the point having the galactic coordinates l, b. The numerical factor 4.74 is a rounded-off value of 4.738. The quantity l_0 is the longitude of the galactic center in the system of galactic coordinates used by Oort. As we will use new galactic coordinates, $l_0 = 0°$. The quantity B is Oort's second constant.

Starting from the first of the Equations (P. VII.17) derive Oort's second equation and show that

$$B = A + \omega_0 \tag{P.VII.20}$$

where ω_0 is the angular velocity of rotation at the Sun's distance from the galactic center.

Solution

Develop the first term on the right side of the first Equation (P. VII.17) using the same approximations as in VII.6. This gives

$$
\begin{aligned}
-(\omega - \omega_0)\,R_0 \cos l &= -(R - R_0)\,\omega'_0 \cdot R_0 \cos l \\
&= + r \cos b \cos l \cdot \omega'_0 \cdot R_0 \cos l \\
&= 2Ar \cos^2 l \cos b = Ar \cos b\,(1 + \cos 2l)
\end{aligned}
\tag{P.VII.21}
$$

where use has been made of the goniometric relation

$$2 \cos^2 l = 1 + \cos 2l. \tag{P.VII.22}$$

In the second term develop $\omega = \omega(R)$ retaining again only the linear term in $R - R_0$; neglect powers of r higher than the first. This gives:

$$\omega r \cos b = [\omega_0 + (R - R_0)\,\omega'_0]\,r \cos b \simeq \omega_0 r \cos b. \tag{P.VII.23}$$

Substitute all this in the first of the Equations (P. VII.17). Then

$$4.738 r\mu_l \cos b - u_0 \sin l + v_0 \cos l = (A \cos 2l + A + \omega_0)\,r \cos b \tag{P.VII.24}$$

or, by dividing by $4.738r$,

$$\mu_l \cos b - \frac{u_0}{4.738r} \sin l + \frac{v_0}{4.738r} \cos l = \left(\frac{A}{4.738} \cos 2l + \frac{B}{4.738} \right) \cos b$$

(P.VII.25)

where we have put

$$A + \omega_0 = B.$$

(P.VII.26)

The expression on the left side of (P. VII.25) is the component of the proper motion in galactic longitude corrected for solar motion.

Oort's second equation follows from (P. VII.25) by averaging.

For a group of stars at a given distance and in the galactic plane ($\cos b = 1$) this equation describes a double wave effect. Notice however that,

First, the double wave effect on the proper motions in galactic longitude is shifted in phase by 45° with respect to the double wave of the radial velocities;

Second, the double wave is shifted, along the vertical axis by $B/4.738$;

Third, at the degree of approximation adopted, the amplitude of the wave does not depend on the distance of the stars considered.

If from an analysis of the radial velocities as well as of the proper motions one has derived the values of the two constants A and B, then from Equation (P. VII.26) one can find the angular velocity at the Sun's distance from the galactic center. The value of B adopted at present is $B = -10$ km/s·kpc, which gives $\omega_0 = -25$ km/s·kpc. The negative sign corresponds to a clockwise rotation of the Galaxy when observed from the north side of the galactic plane.

Notice finally that when effectively using Equation (P. VII.25) the values of the constants A and B must be expressed in $''/a$.

References

[1] Oort, J. H.: 1927, *Bull. Astron. Inst. Neth.* **3**, 275.
[2] Stibbs, D. W. N.: 1956, *Monthly Notices Roy. Astron. Soc.* **116**, 453.
[3] Kraft, R. P. and Schmidt, M.: 1963, *Astrophys. J.* **137**, 249.
[4] Kukarkin, B. V., Parenago, P. P., Yefremov, Y. I., and Holopov, P. N.: 1958, *General Catalogue of Variable Stars*, Academy of Sciences of the U.S.S.R., Moscow.
[5] Camm, G. L.: 1939, *Monthly Notices Roy. Astron. Soc.* **99**, 71.
[6] Gascoigne, S. C. B. and Eggen, O. J.: 1957, *Monthly Notices Roy. Astron. Soc.* **117**, 430.
[7] Johnson, H. L. and Svolopoulos, S. N.: 1961, *Astrophys. J.* **134**, 868.
[8] Feast, M. W. and Shuttleworth, M.: 1965, *Monthly Notices Roy. Astron. Soc.* **130**, 245.
[9] Schmidt, M.: 1965, in A. Blaauw and M. Schmidt (eds.), *Galactic Structure*, Chicago University Press, Chicago, p. 513.

TABLE VII.1

Star	l	$\sin l$	$\cos l$	b	$\sin b$	$\cos b$	ϱ	r	$R - R_0$	ϱ'
X Sgr	1°.2	+0.021	+1.000	+ 0°.2	+0.003	+1.000	−13.7	0.41	−0.41	− 4.4
W Sgr	1 .6	+0.028	+1.000	− 4 .0	−0.070	+0.998	−26.4	0.45	−0.45	−17.6
AP Sgr	8 .1	+0.141	+0.990	− 2 .5	−0.044	+0.999	−18.0	0.88	−0.87	− 7.7
WZ Sgr	12 .1	+0.210	+0.978	− 1 .3	−0.023	+1.000	−10.0	1.75	−1.70	+ 1.2
Y Sgr	12 .8	+0.222	+0.975	− 2 .1	−0.037	+0.999	− 3.8	0.54	−0.53	+ 7.4
AY Sgr	13 .3	+0.230	+0.973	− 2 .4	−0.042	+0.999	−26.5	2.02	−1.95	−15.3
U Sgr	13 .7	+0.237	+0.972	− 4 .5	−0.078	+0.997	+ 3.9	0.64	−0.62	+14.9
V 350 Sgr	13 .8	+0.239	+0.971	− 8 .0	−0.139	+0.990	+ 9.0	0.99	−0.95	+19.5
BB Sgr	14 .7	+0.254	+0.967	− 9 .0	−0.156	+0.988	+ 7.5	0.86	−0.82	+18.0
XX Sgr	15 .0	+0.259	+0.966	− 1 .9	−0.033	+0.999	+ 2.0	1.64	−1.57	+13.6
YZ Sgr	17 .8	+0.306	+0.952	− 7 .1	−0.124	+0.992	+19.3	1.26	−1.18	+30.6
X Sct	19 .0	+0.326	+0.946	− 1 .6	−0.028	+1.000	+ 7.0	1.90	−1.77	+19.2
Y Sct	23 .9	+0.405	+0.914	− 0 .9	−0.016	+1.000	+ 6.5	1.68	−1.51	+19.5
SS Sct	25 .2	+0.426	+0.905	− 1 .8	−0.031	+1.000	−14.0	1.09	−0.97	− 1.0
Z Sct	26 .7	+0.449	+0.893	− 0 .8	−0.014	+1.000	+29.0	2.64	−2.27	+42.3
TY Sct	28 .0	+0.469	+0.883	+ 0 .1	+0.002	+1.000	+ 6.5	2.46	−2.09	+20.1
V 496 Aql	28 .2	+0.473	+0.881	− 7 .1	−0.124	+0.992	+ 5.2	0.99	−0.85	+17.8
RU Sct	28 .2	+0.473	+0.881	− 0 .2	−0.003	+1.000	−12.0	2.02	−1.72	+ 1.6
U Aql	30 .9	+0.514	+0.858	−11 .6	−0.201	+0.980	− 6.8	0.69	−0.57	+ 5.4
V 336 Aql	34 .1	+0.561	+0.828	− 2 .2	−0.038	+0.999	+11.5	2.12	−1.67	+25.4
SZ Aql	35 .6	+0.582	+0.813	− 2 .4	−0.042	+0.999	+ 9.5	2.11	−1.62	+23.5
TT Aql	36 .0	+0.588	+0.809	− 3 .1	−0.054	+0.999	+ 0.5	1.00	−0.79	+14.4
FN Aql	38 .5	+0.623	+0.783	− 3 .1	−0.054	+0.999	+ 6.6	1.63	−1.22	+20.7
η Aql	40 .9	+0.655	+0.756	−13 .1	−0.227	+0.974	−12.9	0.29	−0.21	− 0.2
FM Aql	44 .3	+0.698	+0.716	+ 0 .9	+0.016	+1.000	−10.0	0.96	−0.66	+ 4.9
FF Aql	49 .2	+0.757	+0.653	+ 6 .4	+0.111	+0.994	−17.4	0.39	−0.25	− 1.8
S Sge	55 .2	+0.821	+0.571	− 6 .1	−0.106	+0.994	−10.0	0.68	−0.37	+ 4.2
SV Vul	64 .0	+0.899	+0.438	+ 0 .0	0.000	+1.000	− 0.5	2.07	−0.72	+14.2
SU Cyg	64 .8	+0.905	+0.426	+ 2 .5	+0.044	+0.999	−30.6	0.94	−0.36	−15.6
GH Cyg	66 .2	+0.915	+0.404	− 0 .2	−0.003	+1.000	−19.5	2.53	−0.73	− 4.9
MW Cyg	70 .8	+0.944	+0.329	− 0 .7	−0.012	+1.000	−13.6	1.51	−0.39	+ 0.6
CD Cyg	71 .1	+0.946	+0.324	+ 1 .4	+0.024	+1.000	−10.3	2.44	−0.51	+ 4.1
DT Cyg	76 .6	+0.973	+0.232	−10 .8	−0.187	+0.982	+ 0.6	0.44	−0.09	+12.8
X Cyg	76 .9	+0.974	+0.227	− 4 .3	−0.075	+0.997	+ 9.3	0.95	−0.17	+22.5
VX Cyg	82 .2	+0.991	+0.136	− 3 .5	−0.061	+0.998	−18.0	2.54	−0.02	− 5.3
VY Cyg	82 .9	+0.992	+0.124	− 4 .6	−0.080	+0.997	− 9.8	2.21	−0.03	+ 2.6
TX Cyg	84 .1	+0.995	+0.103	− 2 .0	−0.035	+0.999	−20.7	1.30	−0.05	− 8.1
SZ Cyg	84 .5	+0.995	+0.096	+ 4 .0	+0.070	+0.998	−11.7	2.15	+0.02	+ 1.6
BZ Cyg	84 .8	+0.996	+0.091	+ 1 .4	+0.024	+1.000	−13.2	2.18	+0.04	− 0.3
V 386 Cyg	85 .6	+0.997	+0.077	− 5 .0	−0.087	+0.996	−10.0	1.11	−0.02	+ 2.0
VZ Cyg	91 .5	+1.000	−0.026	− 8 .5	−0.148	+0.989	−16.5	1.95	+0.23	− 5.9
Y Lac	98 .5	+0.989	−0.145	− 4 .0	−0.070	+0.998	−18.0	2.38	+0.61	− 7.9

Table VII.1 (continued)

Star	l	$\sin l$	$\cos l$	b	$\sin b$	$\cos b$	ϱ	r	$R - R_0$	ϱ'
AK Cep	105°.1	+0.965	−0.261	+ 0°.4	+0.007	+1.000	−45.6	3.68	+1.52	−36.3
δ Cep	105 .2	+0.965	−0.262	+ 0 .8	+0.014	+1.000	−17.4	0.26	+0.07	− 8.1
RR Lac	105 .7	+0.963	−0.271	− 2 .0	−0.035	+0.999	−34.5	2.19	+0.80	−25.6
Z Lac	105 .8	+0.962	−0.272	− 1 .6	−0.028	+1.000	−25.0	1.71	+0.59	−16.1
V Lac	106 .5	+0.959	−0.284	− 2 .6	−0.045	+0.999	−20.0	1.77	+0.64	−11.4
X Lac	106 .6	+0.958	−0.286	− 2 .5	−0.044	+0.999	−25.0	1.47	+0.51	−16.4
SW Cas	109 .7	+0.941	−0.337	− 1 .6	−0.028	+1.000	−38.0	2.16	+0.92	−29.9
RS Cas	114 .5	+0.910	−0.415	+ 0 .8	+0.014	+1.000	−24.5	1.72	+0.83	−17.2
RY Cas	115 .3	+0.904	−0.427	− 3 .3	−0.058	+0.998	−70.5	2.90	+1.54	−63.9
DD Cas	116 .8	+0.893	−0.451	+ 0 .5	+0.009	+1.000	−69.5	3.08	+1.72	−62.8
CG Cas	116 .9	+0.892	−0.452	− 1 .3	−0.023	+1.000	−87.0	3.19	+1.79	−80.5
SY Cas	118 .2	+0.881	−0.473	− 4 .1	−0.071	+0.997	−43.0	2.15	+1.17	−37.2
DL Cas	120 .3	+0.863	−0.505	− 2 .5	−0.044	+0.999	−11.0	1.83	+1.04	− 5.5
AP Cas	120 .9	+0.858	−0.514	+ 0 .1	+0.002	+1.000	−44.5	3.78	+2.38	−38.8
XY Cas	122 .8	+0.841	−0.542	− 2 .8	−0.049	+0.999	−42.0	2.20	+1.34	−37.1
VW Cas	124 .6	+0.823	−0.568	− 1 .1	−0.019	+1.000	−58.5	3.24	+2.14	−53.9
BP Cas	125 .4	+0.815	−0.579	+ 2 .9	+0.051	+0.999	−46.5	2.41	+1.56	−41.6
UZ Cas	125 .5	+0.814	−0.581	− 1 .6	−0.028	+1.000	−51.0	3.96	+2.72	−46.7
RW Cas	129 .0	+0.777	−0.629	−- 4 .6	−0.080	+0.997	−64.5	2.83	+1.98	−61.4
BY Cas	129 .6	+0.771	−0.637	− 0 .7	−0.012	+1.000	−45.5	1.45	+0.98	−42.1
VV Cas	130 .4	+0.762	−0.648	− 2 .1	−0.037	+0.999	−50.5	3.72	+2.73	−47.5
VX Per	132 .8	+0.734	−0.679	− 3 .0	−0.052	+0.999	−33.0	2.55	+1.88	−30.7
SU Cas	133 .5	+0.725	−0.688	+ 8 .5	+0.148	+0.989	− 7.8	0.29	+0.20	− 4.3
VY Per	135 .1	+0.706	−0.708	− 1 .6	−0.028	+1.000	−39.5	2.18	+1.65	−37.6
UY Per	135 .9	+0.696	−0.718	− 1 .4	−0.024	+1.000	−59.0	2.40	+1.84	−57.3
RW Cam	144 .9	+0.575	−0.818	+ 3 .8	+0.066	+0.998	−26.0	1.85	+1.56	−26.0
RX Cam	145 .9	+0.561	−0.828	+ 4 .7	+0.082	+0.997	−36.5	0.93	+0.78	−36.6
AS Per	154 .2	+0.435	−0.900	− 1 .0	−0.017	+1.000	−25.5	1.44	+1.31	−28.5
SX Per	159 .0	+0.358	−0.934	− 6 .5	−0.113	+0.994	+ 5.5	3.42	+3.23	+ 0.6
SY Aur	164 .7	+0.264	−0.965	+ 2 .1	+0.037	+0.999	− 2.0	2.75	+2.67	− 7.3
RX Aur	165 .8	+0.245	−0.969	− 1 .3	−0.023	+1.000	−21.3	1.90	+1.85	−27.2
Y Aur	166 .8	+0.228	−0.974	+ 4 .3	+0.075	+0.997	+ 8.5	2.06	+2.01	+ 3.0
RT Aur	183 .2	−0.056	−0.998	+ 8 .9	+0.155	+0.988	+21.0	0.45	+0.44	+12.5
AA Gem	184 .6	−0.080	−0.997	+ 2 .7	+0.047	+0.999	+ 9.5	3.70	+3.69	− 0.1
RZ Gem	187 .7	−0.134	−0.991	− 0 .1	−0.002	+1.000	+ 6.5	2.61	+2.59	− 4.0
AD Gem	193 .2	−0.228	−0.974	+ 7 .7	+0.134	+0.991	+36.0	3.08	+2.99	+25.5
ζ Gem	195 .8	−0.272	−0.962	+11 .9	+0.206	+0.979	+ 6.8	0.33	+0.31	− 3.4
RS Ori	196 .6	−0.286	−0.958	+ 0 .3	+0.005	+1.000	+40.5	1.90	+1.83	+28.5
T Mon	203 .6	−0.400	−0.916	− 2 .6	−0.045	+0.999	+32.5	1.09	+1.01	+19.2
TX Mon	214 .1	−0.561	−0.828	− 0 .8	−0.014	+1.000	+51.0	5.25	+4.65	+36.7

Table VII.1 (continued)

Star	l	$\sin l$	$\cos l$	b	$\sin b$	$\cos b$	ϱ	r	$R - R_0$	ϱ'
RY CMa	226°.0	− 0.719	− 0.695	+ 0°.3	+ 0.005	+ 1.000	+ 37.5	1.33	+ 0.97	+ 22.7
VW Pup	235 .3	− 0.822	− 0.569	− 0 .6	− 0.010	+ 1.000	+ 24.0	4.08	+ 2.77	+ 8.9
X Pup	236 .1	− 0.830	− 0.558	− 0 .8	− 0.014	+ 1.000	+ 64.0	2.95	+ 1.90	+ 48.9
SS CMa	239 .2	− 0.859	− 0.512	− 4 .2	− 0.073	+ 0.997	+ 60.0	3.81	+ 2.38	+ 44.6
WX Pup	241 .5	− 0.879	− 0.477	− 1 .4	− 0.024	+ 1.000	+ 53.4	2.68	+ 1.52	+ 38.4
WZ Pup	241 .8	− 0.881	− 0.473	+ 3 .3	+ 0.058	+ 0.998	+ 64.0	4.15	+ 2.50	+ 49.6
AD Pup	241 .9	− 0.882	− 0.471	− 0 .0	0.000	+ 1.000	+ 67.5	4.61	+ 2.83	+ 52.7
VZ Pup	243 .4	− 0.894	− 0.448	− 3 .3	− 0.058	+ 0.998	+ 49.0	4.79	+ 2.87	+ 33.9
AQ Pup	246 .1	− 0.914	− 0.405	+ 0 .1	+ 0.002	+ 1.000	+ 59.8	2.82	+ 1.44	+ 45.2
AT Pup	254 .3	− 0.963	− 0.271	− 1 .6	− 0.028	+ 1.000	+ 28.8	1.72	+ 0.60	+ 14.6
AP Pup	255 .5	− 0.968	− 0.250	− 5 .7	− 0.099	+ 0.995	+ 17.9	1.03	+ 0.30	+ 3.4
AX Vel	263 .2	− 0.993	− 0.118	− 7 .7	− 0.134	+ 0.991	+ 24.2	1.07	+ 0.18	+ 10.4
T Vel	265 .3	− 0.997	− 0.082	− 4 .1	− 0.071	+ 0.997	+ 8.0	1.15	+ 0.16	− 5.2
SX Vel	265 .5	− 0.997	− 0.078	− 2 .2	− 0.038	+ 0.999	+ 30.9	2.34	+ 0.45	+ 18.0
β Dor	271 .7	− 1.000	+ 0.030	− 32 .8	− 0.542	+ 0.841	+ 8.1	0.33	0.00	− 5.6
BG Vel	271 .9	− 0.999	+ 0.033	− 2 .5	− 0.044	+ 0.999	+ 10.9	0.93	+ 0.01	− 1.1
V Vel	276 .6	− 0.993	+ 0.115	− 4 .2	− 0.073	+ 0.997	− 28.7	1.15	− 0.07	− 40.1
l Car	283 .2	− 0.974	+ 0.228	− 7 .0	− 0.122	+ 0.993	+ 1.4	0.40	− 0.08	− 9.0
UX Car	284 .8	− 0.967	+ 0.255	+ 0 .2	+ 0.003	+ 1.000	+ 10.4	1.60	− 0.28	+ 1.1
Y Car	285 .7	− 0.963	+ 0.271	− 0 .3	− 0.005	+ 1.000	− 5.8	1.54	− 0.30	− 15.0
SV Vel	286 .0	− 0.961	+ 0.276	+ 2 .4	+ 0.042	+ 0.999	+ 4.5	2.47	− 0.38	− 4.2
VY Car	286 .5	− 0.959	+ 0.284	+ 1 .2	+ 0.021	+ 1.000	− 1.2	1.61	− 0.33	− 10.0
ER Car	290 .1	− 0.939	+ 0.344	+ 1 .5	+ 0.026	+ 1.000	− 18.1	1.02	− 0.30	− 26.1
IT Car	291 .5	− 0.930	+ 0.367	− 1 .1	− 0.019	+ 1.000	− 14.0	1.60	− 0.47	− 22.0
AZ Cen	292 .8	− 0.922	+ 0.388	− 0 .2	− 0.003	+ 1.000	− 12.3	1.72	− 0.53	− 19.9
UZ Cen	294 .9	− 0.907	+ 0.421	− 0 .9	− 0.016	+ 1.000	− 5.1	1.66	− 0.58	− 12.3
RT Mus	296 .5	− 0.894	+ 0.446	− 5 .3	− 0.092	+ 0.996	− 1.8	1.45	− 0.56	− 9.1
T Cru	299 .4	− 0.871	+ 0.491	+ 0 .4	+ 0.007	+ 1.000	− 6.0	0.80	− 0.37	− 12.0
R Cru	299 .7	− 0.869	+ 0.495	+ 1 .1	+ 0.019	+ 1.000	− 13.5	0.94	− 0.43	− 19.3
S Mus	299 .7	− 0.869	+ 0.495	− 7 .5	− 0.131	+ 0.991	+ 11.8	0.96	− 0.44	+ 5.0
AG Cru	301 .7	− 0.851	+ 0.525	+ 3 .1	+ 0.054	+ 0.999	− 4.5	1.42	− 0.67	− 9.6
R Mus	301 .9	− 0.849	+ 0.528	− 6 .6	− 0.115	+ 0.993	+ 3.8	0.94	− 0.46	− 2.4
X Cru	302 .3	− 0.845	+ 0.534	+ 3 .7	+ 0.065	+ 0.998	− 25.0	1.59	− 0.75	− 29.9
S Cru	303 .3	− 0.836	+ 0.549	+ 4 .4	+ 0.077	+ 0.997	− 6.6	0.76	− 0.40	− 11.1
XX Cen	309 .5	− 0.772	+ 0.636	+ 4 .6	+ 0.080	+ 0.997	− 15.7	1.79	− 1.03	− 18.7
V Cen	316 .4	− 0.690	+ 0.724	+ 3 .3	+ 0.058	+ 0.998	− 22.3	0.82	− 0.58	− 23.7
R TrA	317 .0	− 0.682	+ 0.731	− 7 .8	− 0.136	+ 0.991	− 9.1	0.66	− 0.47	− 11.6
S TrA	322 .1	− 0.614	+ 0.789	− 8 .2	− 0.143	+ 0.990	+ 6.6	0.88	− 0.67	+ 5.3
S Nor	327 .8	− 0.532	+ 0.846	− 5 .4	− 0.094	+ 0.996	+ 3.3	0.90	− 0.75	+ 3.9
RV Sco	350 .4	− 0.167	+ 0.986	+ 5 .7	+ 0.099	+ 0.995	− 18.3	0.84	− 0.82	− 10.8
V 482 Sco	354 .3	− 0.099	+ 0.995	+ 0 .2	+ 0.003	+ 1.000	+ 11.1	1.03	− 1.02	+ 18.9
RY Sco	356 .5	− 0.061	+ 0.998	− 3 .4	− 0.059	+ 0.998	− 18.4	1.51	− 1.50	− 10.6
BF Oph	357 .1	− 0.051	+ 0.999	+ 8 .6	+ 0.150	+ 0.989	− 31.4	0.85	− 0.84	− 22.1
V 500 Sco	359 .0	− 0.017	+ 1.000	− 1 .4	− 0.024	+ 1.000	− 13.8	1.64	− 1.64	− 5.2

THE DETERMINATION OF THE GALACTIC
ORBIT OF A STAR

1. In the preceding exercises we have investigated some of the major problems concerning the distribution and the motion of the stars in our Galaxy. In either case the emphasis has been on the statistical point of view and, so far as the problems of motion are concerned, it is only the kinematical aspect which has been covered. This last exercise will be devoted to a problem in stellar dynamics. Moreover we shall radically change our standpoint and focus our attention on the individual rather than on a large community of stars. In fact we shall show how the galactic orbit of a star can be computed and propose to the reader that he should determine the orbit of a real star.

The question to be dealt with clearly bears some resemblance to problems of planetary motion considered in classical celestial mechanics. This is especially true as to the mathematics used. However there are also deep-rooted differences between the two problems. In fact, ever since the discovery of the existence of different types of stellar populations in our Galaxy, there has been a strong tendency to explain the differences between them by evolutionary effects, or, to be more precise, by differences in origin, age, and chemical constitution. This idea has been put forward in the early forties by B. V. Kukarkin, and developed since by many investigators. One of its most explored aspects is the problem of the correlations between the parameters of the orbit of a star and its physical properties (see, e.g., [1] where several orbits of stars have been given). More recently an extremely interesting and promising possibility has been opened by the work of B. Strömgren and his associates on multi-color narrow-band photometry of stars (see, e.g. [2] and [3]). It seems that, at present, the ages of certain classes of stars can be determined so accurately that by computing their motion backwards in time we can hope to locate their place of origin. This is a reason why orbit computations have wider implications in stellar astronomy than in celestial mechanics.

An earlier comprehensive investigation on plane galactic orbits which also covers the empirical aspect of the problem and which contains data on, and graphs of, the orbits of ten stars, is that by I. Torgard [4]. An exhaustive study of the more general problem of three-dimensional stellar orbits from the standpoint of theoretical stellar dynamics has been made by A. Ollongren (see [5] for a general account). In the present exercise we shall limit ourselves to plane orbit determination considering this as a purely mechanical problem.

2. Let us first state our basic assumption as well as the fundamental equation of our problem. As far as the first point is concerned we shall admit that the force which determines the motions of the stars is of gravitational character. Moreover we shall

(a) neglect the irregularities in the distribution of the stars and of all other matter in the Galaxy, and substitute for the real distribution of the attracting masses a smoothed-out one, symmetric to the galactic plane and possessing rotational symmetry around the galactic axis (which is taken to be perpendicular to the galactic plane and drawn through the galactic center);

(b) neglect the effect of stellar encounters, i.e. assume that the orbit of any given star is determined with sufficient accuracy by a mass distribution as defined above, which means that perturbations caused by the close passages of other stars will be considered as vanishingly small.

This last point is discussed in all standard works on stellar dynamics. The most thorough investigation is that by S. Chandrasekhar in his classical *Principles of Stellar Dynamics* [15].

As in our investigations we will make use of Lagrange's equations in generalised coordinates, we first have to choose coordinates adapted to our problem. In view of the symmetry properties of our stellar system we obviously are led to adopt cylindrical coordinates, see Figure VIII.1.

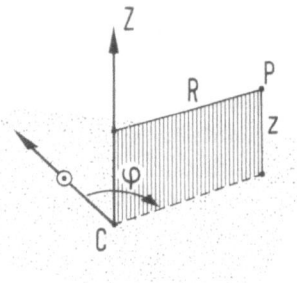

Fig. VIII.1.

As the z-axis we take the galactic axis defined above, with the zero point in the galactic center and pointing towards the North Galactic Pole. The distance from this axis will be denoted by R, the coordinate measured along it by z. Obviously $|z|$ is the distance from the galactic plane. Assuming that the Sun is in the galactic plane, we shall draw an axis from the galactic center towards the position of the Sun at a given moment. The (third) cylindrical coordinate will be the azimuthal angle reckoned from this axis in the direction of galactic rotation, i.e. clockwise as seen from the north face of the galactic plane.

Consider first the general case of three-dimensional (or space) motion. It can be shown (see Problem VIII.1) that the differential equations of the motion of the star will be

$$\ddot{R} - R\dot{\varphi}^2 = F_R, \qquad \frac{\mathrm{d}}{\mathrm{d}t}(R^2\dot{\varphi}) = 0, \qquad \ddot{z} = F_z. \tag{VIII.1}$$

where F_R and F_z are the components of the force, per unit mass, which correspond to the coordinates R and z respectively.

Moreover, it is not difficult to prove that any star which at some initial moment of time is in the galactic plane and at this moment has no velocity component along the z-axis, will for ever remain in that plane, i.e. describe a plane orbit. For such a star the differential Equations (VII.1) reduce to

$$\ddot{R} - R\dot{\varphi}^2 = F_R, \qquad \frac{d}{dt}(R^2\dot{\varphi}) = 0 \qquad \text{(VIII.2)}$$

(see Problem VIII.2).

Obviously there will not be so many stars with z-components of the velocity exactly equal to zero. However it can be expected, and in fact proved, that for stars with low z-velocities, the plane orbit derived under the assumption $z=0$ will represent a fair approximation to the real orbit which can be used to start an iterative process leading to the determination of the true, space orbit of the star [6].

Let us now consider the plane orbits more closely. We immediately can write down the integral of the second of Equations (VIII.2) which is

$$R^2\dot{\varphi} = h. \qquad \text{(VIII.3)}$$

This is, in fact, the integral of area well known from the two-body problem. The quantity h is an integration constant. It can be determined from the initial condition, i.e. from the position and the velocity of the star at the moment $t=t_0$.

There also exists a second integral which is

$$\tfrac{1}{2}(\dot{R}^2 + h^2/R^2) - \Phi(R) = H, \quad \text{or} \quad \tfrac{1}{2}(\dot{R}^2 + R^2\dot{\varphi}^2) - \Phi(R) = H, \qquad \text{(VIII.4)}$$

The first term on the left side is the kinetic energy, whereas $-\Phi(R)$ is the potential energy. Then $\Phi(R)$ is the gravitational potential in the galactic plane, at the distance R from the center. As known from elementary mechanics F_R and $\Phi(R)$ are interrelated by

$$F_R = d\Phi/dR \quad \text{and} \quad \Phi = \int F_R \, dR + C \qquad \text{(VIII.5)}$$

where the constant C shall be chosen in such a way that Φ becomes zero at infinity. H is another integration constant which, like h, has to be determined from the initial conditions. This second integral is, in fact, the energy integral (see Problem VIII.3).

From these two integrals the differential equation of the orbit can be derived as follows. From (VIII.4) we have

$$\dot{R}^2 = [2(\Phi + H) R^2 - h^2]/R^2, \qquad \text{(VIII.6)}$$

or

$$\frac{dR}{dt} = \frac{1}{R}[2(\Phi + H) R^2 - h^2]^{1/2}, \qquad \text{(VIII.7)}$$

whereas (VIII.3) gives

$$1/\dot{\varphi} = dt/d\varphi = R^2/h.$$ (VIII.8)

By combining the last two equations we get

$$\frac{dR}{d\varphi} = R \left[\frac{2}{h^2} \left(\Phi + H \right) R^2 - 1 \right]^{1/2},$$ (VIII.9)

which is the differential equation of the orbit. The sign of the square root has to be chosen in accordance with the initial condition (see Section 5).

3. In order to determine the orbit of a star we obviously must know the force field in which it is moving, or, what amounts to the same, the function $\Phi(R)$.

Now it is just at this point that our account, which so far has followed the same lines as in the two-body problem, must take a different direction. In fact, in the two-body problem the force is a known function of R given by Newton's law of universal attraction. In the present case the force law $F(R)$ is not a general law but instead a specific property of our stellar system which has to be determined.

The problem of the determination of either $F(R)$ or $\Phi(R)$ can be, and effectively has been, approached in different ways. References to older work can be found in Trumpler's and Weaver's monograph [7] or in Ogorodnikov's book [8]. In her paper already cited I. Torgard gives a detailed account of the derivation of the force law in the galactic plane which she has used to compute a series of plane galactic orbits. The most recent derivation of $F(R)$ is that by G. Contopoulos and B. Strömgren. It has served as a basis for their very comprehensive Tables of plane galactic orbits [9], to which the reader is referred for an account of the work of these authors as well as for a very complete bibliography on the galactic orbits in general.

The method used by Torgard as well as by Contopoulos and Strömgren is in fact that proposed by G. L. Camm, an outline of which is given at the end of the long quotation from Camm in Exercise VII, Section 7. Consider a star which, moving with the appropriate velocity, describes a circular orbit of radius R in the galactic plane. Then R=const, so that $\ddot{R}$=0, and the first of the Equations (VIII.1) becomes

$$R\dot{\varphi}^2 = - F_R = - F(R).$$ (VIII.10)

Or, remembering that $\dot{\varphi} = \omega = \omega(R)$ is the angular velocity of the star in its circular orbit around the galactic center

$$F_R = F(R) = - R \cdot \omega (R)^2.$$ (VIII.11)

The derivation of the angular velocity curve, i.e. of the relation $\omega = \omega(R)$ from an analysis of the radial velocities of the classical cepheids has been discussed at the end of the preceding exercise. Now, if we assume, with Camm, that these angular velocities "correspond to circular orbits described under gravitational attraction of the whole Galaxy (then) it is possible to find the variation of the force with the galactic radius", that is as a function of R, by making use of Equation (VIII.11).

It is important to realise that this method needs some comment and justification. In fact, we shall not forget that the equations of the theory of galactic rotation apply to the motions of the centroids, whereas we now are considering the motion of mass points. It would therefore be necessary to investigate and to clarify the problem of the relation between the respective velocities of a star moving in a circular orbit of radius R in the galactic plane, and that of a centroid, in the galactic plane, and at the same distance from the center. This problem is related to the phenomenon of the asymmetry of the stellar motions. Unfortunately it cannot be discussed in this exercise. Once again we are compelled to refer the more interested reader to some treatise in stellar dynamics, e.g. [8] for a general discussion, as well as to an older but very important study by P. P. Parenago [10] and J. Delhaye's article [11].

In the publications by Torgard as well as by Contopoulos and Strömgren referred to above, the angular velocities at a series of distances from the galactic center have been used to derive, by the method of least squares, an empirical relation between the force (per unit mass) and the distance R. For distances larger than the Sun's distance from the galactic center radial velocities of the classical cepheids have been used. For distances less than the Sun's distance from the center the determination is based on radio-astronomical measurement of the 21 cm H I line. The relation found by the last two authors is

$$F_R = 73340/R^2 - 1581.8 + 3442.03\,R - 402.621\,R^2 + 12.9402\,R^3.$$

As the angular velocities are expressed in km/s per kpc, and the distances measured in kpc, the unit for the force (per unit mass) will be $(\text{km/s})^2$ per kpc.

4. It is very interesting that, for a wide range of the galactocentric distances R, the velocity field from which the force law is derived, can quite exactly be described by the formula

$$V = \omega R = -k_1 R/(1 + k_2 R^2). \tag{VIII.12}$$

This formula has, moreover, the advantage that it follows from certain general relations of stellar dynamics (see, e.g. [8], [12]). Making use of Equation (VIII.11) we immediately get from (VIII.12) the following expression for the force

$$F_R = -k_1^2 R/(1 + k_2 R^2)^2 \tag{VIII.13}$$

and further

$$\Phi(R) = \Phi_0/(1 + k_2 R^2), \qquad \Phi_0 = k_1^2/2k_2. \tag{VIII.14}$$

Now the values of the constants k_1 and k_2 can be determined from Oort's constants A and B and the Sun's distance from the galactic center, R_0.

From (VIII.12) we immediately get

$$\omega = -k_1/(1 + k_2 R^2), \quad \text{or} \quad k_1 = -(1 + k_2 R^2)\,\omega. \tag{VIII.15}$$

Putting $R = R_0$ and making use of the relation $\omega(R_0) = B - A$ we have

$$k_1 = -(1 + k_2 R_0^2)(B - A). \tag{VIII.16}$$

By differentiating Equation (VIII.12) with respect to R we have further

$$\frac{dV}{dR} = -k_1 \frac{1 - k_2 R^2}{(1 + k_2 R^2)^2} = \frac{d}{dR}(\omega R) = R\frac{d\omega}{dR} + \omega. \tag{VIII.17}$$

Putting again $R = R_0$ and making use of the relations

$$R_0 \frac{d\omega}{dR_0} = 2A, \qquad \omega_0 = B - A \qquad \text{(VIII.18)}$$

derived in the theory of galactic rotation (Exercise VII, Problem VII.3) we have

$$B + A = -k_1 \frac{1 - k_2 R_0^2}{(1 + k_2 R_0^2)^2} = \frac{1 - k_2 R_0^2}{1 + k_2 R_0^2} (B - A) \qquad \text{(VIII.19)}$$

which after some transformation gives

$$k_2 = -\frac{1}{R_0^2} \cdot \frac{A}{B}. \qquad \text{(VIII.20)}$$

Substituting the values at present adopted for A, B, and R_0, which are

$$A = + 15 \text{ km/s·kpc}, \qquad B = -10 \text{ km/s·kpc}, \qquad R_0 = 10 \text{ kpc},$$

we finally get

$$V = 62.5 \, R/(1 + 0.015 \, R^2). \qquad \text{(VIII.21)}$$

Table VII.1 contains, in the second row, the values of V, in km/s, adopted by Contopoulos and Strömgren when deriving their force law; and, in the third row, the intensity F of the force per unit mass, in (km/s)^2 per kpc according to this law. The accuracy of F is of the order of ± 300. The distance R from the galactic center is given in kpc.

TABLE VIII.1

R	5	6	7	8	9	10	11	12	13	14
V	225	238	247	252	253	250	244	235	228	224
F	10110	9410	8720	7960	7120	6250	5390	4620	3990	3570

Show that between $R = 5$ kpc and $R = 14$ kpc the values for V computed by the formula (VIII.21) do not differ by more than 6 km/s from the tabular values, and that for the force the difference scarcely exceeds the limit of ± 300.

We can therefore expect that between $R = 5$ kpc and $r = 14$ kpc the force F_R as well as the potential $\Phi(R)$ will be given quite well by the formulae (VIII.13) and (VIII.14) respectively. This last expression for the potential has been proposed and extensively used by P. P. Parenago, a leading Soviet authority in the field of galactic research [13].

Verify that with the values of A, B, and R_0, adopted at present the expression for the potential will be

$$\Phi(R) = 868 \times 10^4/(66.67 + R^2). \qquad \text{(VIII.22)}$$

It is this formula that we shall adopt for the potential.

By substituting (VIII.22) into the general differential equation of the orbit, Equation (VIII.9), we get

$$\frac{dR}{d\varphi} = R \left[\frac{2}{h^2} \left(\frac{868 \times 10^4}{66.67 + R^2} + H \right) R^2 - 1 \right]^{1/2}. \qquad \text{(VIII.23)}$$

This is, then, the differential equation of the orbit of a star moving in the galactic plane under the influence of a gravitational force derivable from the potential (VIII.22).

Of course, the form of the orbit which a given star will describe, that is the exact form of the function $R = R(\varphi)$ will depend on the numerical values of the integration constants h and H. These are, in turn, determined by the initial conditions which are the coordinates of the point in space occupied by the star at a given moment, and its velocity in this point. In other words if, e.g., we know the present position and velocity of a star, then by integrating Equation (VIII.23) we can determine its orbit. This is what we will do.

5. In the preceding sections we have reviewed the general theory of plane orbits determinations. Now we will go over to the effective computation of the orbit of a given star. First we shall determine the integration constants h and H.

As an example we shall take the well-known high velocity star 61 Cygni. It was one of the first stars to have its trigonometric parallax determined (Bessel, 1837). Thanks to a long series or proper motion determinations, to first-class radial velocity measurements, as well as to the fact that its distance from the Sun has been quite well determined, the basic observational data we need are known with a high accuracy.

For the reader's convenience we give here the equatorial coordinates of the star which are

61 Cygni $\alpha = 20$ h 04 m 39.8 s $\delta = +38°29'59''$,   1950.0,

and add some of its most currently used designations

Yale Bright Star Catalogue, BS 8085 (=HR 8085)
Boss's General Catalogue, GC 29509
The Henry Draper Catalogue, HD 201091.

In this section we shall see how the integration constants h and H can be determined. The relevant equations are (VIII.3) and (VIII.4) which we shall write as follows

$$h = R_0^2 \dot\varphi_0, \qquad H = \tfrac{1}{2}(\dot R_0^2 + R_0^2 \dot\varphi_0^2) - \Phi(R_0). \qquad (VIII.24)$$

The subscript zero indicates that the values of the coordinates and of their derivatives refer to the initial moment $t = t_0$.

As the initial moment we shall take the present time. Imprecise as it may seem, such a definition of the initial moment fully meets our needs. In fact, due to the very long intervals of time involved a variation of the initial moment by as much as one century will not appreciably affect the results.

When defining the system of cylindrical coordinates used we have adopted for the axis from which the azimuthal angle φ is reckoned the axis pointing from the galactic center towards the position of the Sun at a given moment. Now we specify that this moment is in fact the present time. Let us moreover introduce a system of U, V, W, axes analogous to that extensively used in the Exercises IV and V, but having as origin the (fixed) position of the Sun at the present time. Notice that both frames, the cylindrical as well as the other one just introduced, are fixed in space, and remember

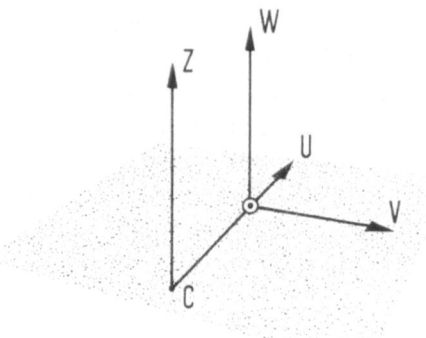

Fig. VIII.2.

that the system of U, V, W, axes previously used was moving with the Sun. This is a very important point deserving special attention (Figure VIII.2).

In order to compute the integration constants h and H we must know the values of the coordinates as well as that of their derivatives appearing on the right sides of Equations (VIII.24).

As to the coordinates which define the initial position only R_0 appears explicitly in the Equations (VIII.24). Now in the *Yale Bright Star Catalogue* we find that the parallax of 61 Cygni is 0.292″ which corresponds to a distance of about 3 parsec. This is a very small quantity compared to other distances appearing in our problem. Therefore it can be neglected. In other words we shall assume that the initial position of 61 Cygni coincides with the present position of the Sun. This means that for the initial values of the coordinates we will adopt

$$R_0 = R(t_0) = 10 \text{ kpc}, \quad \varphi_0 = \varphi(t_0) = 0 \text{ radian}.$$

The determination of the initial velocity is a more difficult problem.

In Eggen's *Catalogue of Space Velocities* [14] we find for the components of the space velocity of 61 Cygni, in the usual U, V, W, system of axes, the following values

$$U = +92.3, \quad V = -53.2, \quad W = -8.5 \text{ km/s}.$$

Now the important point is that we can safely neglect the effect of the displacement of the origin between the epoch to which these values refer and the present time; however we can by no means ignore the fact that the values quoted above refer to a frame moving with the Sun, whereas for the computation of the integration constants we need velocity components referred to a fixed frame. In other words, before using them, we must correct the catalogue values of the velocity components for the effect of the solar motion relative to the frame at rest.

This reduction consists of two steps. First one takes a system of axes analogous to the usual U, V, W system, but having as its origin a fictitious star which describes a circular orbit in the galactic plane at the Sun's distance from the center. The position of this star has to coincide with the Sun's position at $t = t_0$. This frame defines the so-called Local Standard of Rest. According to J. Delhaye [11] the components of the solar velocity with respect to the Local Standard of Rest are

$$U_{\odot LSR} = -9, \qquad V_{\odot LSR} = +12, \qquad W_{\odot LSR} = +7, \quad km/s.$$

The components of the velocity of 61 Cygni relative to the Local Standard of Rest will therefore be

$$U_{*LSR} = +83.3, \qquad V_{*LSR} = -41.2, \qquad W_{*LSR} = -1.5, \quad km/s.$$

In a second step allowance must be made for the motion of the Local Standard of Rest relative to the fixed frame. For the absolute value of the velocity of the fictitious star one adopts, according to the theory of galactic rotation

$$R_0 \cdot \omega_0 = R_0 (B - A) = 250 \text{ km/s},$$

where, in view of the orientation of the coordinate system used here, we have omitted the negative sign. As at the moment $t = t_0$ the position of the fictitious star coincides with the present position of the Sun, we finally get for the components of the velocity of 61 Cygni relative to the fixed system of U, V, W, axes

$$U_{*fs} = +83.8, \qquad V_{*fs} = +208.8, \qquad W_{*fs} = -1.5, \quad km/s.$$

Notice that the last component, which is identical to $\dot{z}(t_0)$ is small compared to the other components; we shall neglect it and assume it to be zero.

Now we obviously have $R_0 \dot{\varphi}_0 = V_{*fs}$, so that the equation for h can be written as follows

$$h = R_0^2 \dot{\varphi}_0 = R_0 \cdot V_{*fs} \tag{VIII.25}$$

which gives

$$h = 2088 \text{ kpc} \cdot \text{km/s}.$$

On the other hand $\dot{R}_0 = U_{*fs}$ so that the expression for H becomes

$$H = \tfrac{1}{2}(U_{*fs}^2 + V_{*fs}^2) - \Phi(R_0). \tag{VIII.26}$$

This gives for H

$$H = -26815 \text{ (km/s)}^2.$$

By substituting these values into (VIII.23) one easily gets

$$\frac{dR}{d\varphi} = R \left[\frac{3.982R^2}{66.67 + R^2} - 0.0123R^2 - 1 \right]^{1/2}. \tag{VIII.27}$$

Notice that of the two possible values of the square root it is the positive one which shall be adopted, as the star is moving outwards ($U_{*fs} = +83.3$ km/s > 0) and in the direction of galactic rotation ($V_{*fs} = +208.8$ km/s > 0), so that $dR/d\varphi > 0$. Notice, further, that the galactocentric distance R of the star oscillates between the extreme values R_{max} and R_{min} which are given by the condition $dR/d\varphi = 0$. The point of

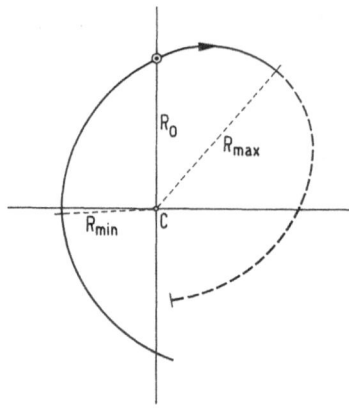

Fig. VIII.3.

closest approach to the galactic center ($R = R_{min}$) is called the pericenter or the peri-galacticum of the star's orbit; the point where R attains its maximal value is the apocenter or apogalacticum. For the orbit of 61 Cygni we have

$$R_{min} = 6.3160, \qquad R_{max} = 11.6567, \quad \text{kpc}.$$

Verify these as well as all the other numerical values given in this Section.

Convince yourself

(a) that the star is now on that part of its orbit which corresponds to the arc between the pericenter and the apocenter;

(b) that, as in this last point $dR/d\varphi = 0$ and thereafter becomes negative, the arc of the orbit between the apocenter and the next following pericenter will be defined by a differential equation of the same form as (VIII.27), but with a negative sign for the square root;

(c) that, for this reason, the arc of the orbit between two consecutive pericenters will be symmetric about the radius $R = R_{max}$ lying between them, so that by placing side by side the arcs between two consecutive extreme values of R, the whole orbit can be constructed.

In the following sections we shall show how the differential Equation (VIII.23) can be numerically solved.

6. To integrate numerically a differential equation means to construct a table giving the values of the unknown function for equidistant values of the independent variable (of the so-called argument). The interval between two consecutive values of the argument is called the integration step. Obviously the solution must satisfy the differential equation as well as the initial conditions.

The Tables VIII.4 and VIII.5 (p. 130) give the numerical solution of the differential Equation (VIII.27) from $\varphi = 0.00$ and $R = 10.0000$ to $\varphi = -1.55$ rad. and

$R=6.3190$, and from the same starting point up to $\varphi = +0.65$ and $R=11.6553$. Taken together they almost exactly cover the arc of the orbit of 61 Cygni between the last pericenter and the next following apocenter.

In general a numerical solution does not give any information about the behavior of the function beyond the limits of the integration table. This disadvantage is however sometimes offset by the fact that a numerical solution can be derived under much wider conditions than an analytical one. Moreover, in the present case, and due to the symmetry discussed at the end of the last Section, we in fact can get a clear idea of the whole orbit even when integrating only between the limits indicated above.

There are many numerical methods for solving differential equations. In the present exercise we shall make use of a method due to the great British 19th century astronomer J. C. Adams, well known for his independent discovery of the planet Neptune. Adams' method has been developed by the outstanding Russian mathematician A. N. Krylov and extended to systems of second-order differential equations by the Norwegian mathematician C. Störmer in connection with his work on solar particles producing the aurorae. It is probably the method most suited for smaller desk computers.

We shall now give a brief account of Adams' method and at the same time explain how our integration tables have been constructed. It is expected that the reader will effectively repeat the calculations indicated in order to get a clear idea of how the method works. It must be stressed, however, that the precision we have adopted in our computations is much too high from an astronomical standpoint (why?). This has been done purposely, so that the reader may independently calculate the solution with the precision really needed and at the same time have a check of his own results. Let us add that the methods of numerical integration represent a very powerful tool in many branches of theoretical astronomical research.

Consider now quite generally the first-order differential equation

$$\frac{dx}{dt} = F(x, t) \tag{VIII.28}$$

where x and t are the function and the argument respectively, and where $F(x, t)$ is a given function of both of them. Let the initial conditions be

$$t = t_0, \quad x = x_0; \quad \left(\frac{dx}{dt}\right)_0 = F(x_0, t_0).$$

We are asked to compute a series of values $x = x_k$ of the function, which correspond to equidistant values t_k of the independent variable. Denote by w the integration step and put

$$w \frac{dx}{dt} = wF(x(t), t) = f(t) \tag{VIII.29}$$

so that

$$w \left(\frac{dx}{dt}\right)_k = wF(x_k, t_k) = f(t_k) = f_k. \tag{VIII.30}$$

TABLE VIII.2

t	x	Δ	f	f^1	f^2	f^3	...
...	...	...	...	...	...	...	...
t_{k-3}	x_{k-3}		f_{k-3}		$f^2{}_{k-3}$		
		$\Delta_{k-5/2}$		$f^1{}_{k-5/2}$		$f^3{}_{k-5/2}$	
t_{k-2}	x_{k-2}		f_{k-2}		$f^2{}_{k-2}$		...
		$\Delta_{k-3/2}$		$f^1{}_{k-3/2}$		$f^3{}_{k-3/2}$	
t_{k-1}	x_{k-1}		f_{k-1}		$f^2{}_{k-1}$		
		$\Delta_{k-1/2}$		$f^1{}_{k-1/2}$			
t_k	x_k		f_k				

The choice of the value for w will be discussed later. In any case w will be considered as a known quantity.

For sake of simplicity suppose first that by some means we already have computed a series of values of x up to x_k inclusive. They will be arranged in a table like Table VIII.2. The first and the second column contain respectively the consecutive values of the argument and of the function. The quantities standing in the column Δ are the increments to be algebraically added to any of the values of x just preceding this increment, in order to get the next following value of x. In other words we have quite generally

$$x_{j+1} = x_j + \Delta_{j+1/2}, \quad \text{so that} \quad \Delta_{j+1/2} = x_{j+1} - x_j. \qquad \text{(VIII.31)}$$

which relations can be considered as the definition of the increment $\Delta_{j+1/2}$.

The fourth column contains the values of $f_k = w \cdot F(x_k, t_k)$. The quantities appearing in the columns to the right of this one are the differences of the first, second, third, ..., and so on, order of f_k. They are defined (but not computed!) just in the same way as the increments. We have, e.g.

$$f^1_{k-1/2} = f_k - f_{k-1}; \quad f^2_{k-1} = f^1_{k-1/2} - f^1_{k-3/2}; \quad f^3_{k-3/2} = f^2_{k-1} - f^2_{k-2}; \ldots$$

Compare the schematical Table VIII.2 to any of the integration tables in order to get a clear insight into the meaning of the quantities they contain and the way in which they are arranged.

According to our assumption the integration has been pushed up to the point t_k, x_k. The question is: How can we compute the next value of x?

Theoretically the answer is quite simple. In fact x_{k+1} is the value of x for $t = t_{k+1} = t_k + w$. Therefore, by developing $x(t)$ around $t = t_k$ we have

$$x_{k+1} = x(t_k + w) = x(t_k) + \frac{w}{1!}\left(\frac{dx}{dt}\right)_k + \frac{w^2}{2!}\left(\frac{d^2x}{dt^2}\right)_k + \frac{w^3}{3!}\left(\frac{d^3x}{dt^3}\right)_k + \ldots$$
$$\text{(VIII.32)}$$

or, by virtue of (VIII.30) and (VIII.31)

$$\Delta_{k+1/2} = f_k + \frac{w}{2}\left(\frac{df}{dt}\right)_k + \frac{w^2}{6}\left(\frac{d^2f}{dt^2}\right)_k + \ldots. \qquad \text{(VIII.33)}$$

Now the successive derivatives of f standing on the right side of (VIII.33) can be computed from the differences lying on the last ascending diagonal of our schematical Table VIII.2. In fact it can be shown that (see Appendix II):

$$w\left(\frac{\mathrm{d}f}{\mathrm{d}t}\right)_k = f_{k-1/2}^1 + \tfrac{1}{2} f_{k-1}^2 + \tfrac{1}{3} f_{k-3/2}^3 + \tfrac{1}{4} f_{k-2}^4 + \tfrac{1}{5} f_{k-5/2}^5 + \cdots$$

$$w^2\left(\frac{\mathrm{d}^2 f}{\mathrm{d}t^2}\right)_k = \qquad f_{k-1}^2 + \quad f_{k-3/2}^3 + \tfrac{11}{12} f_{k-2}^4 + \tfrac{5}{6} f_{k-5/2}^5 + \cdots$$

$$w^3\left(\frac{\mathrm{d}^3 f}{\mathrm{d}t^3}\right)_k = \qquad\qquad f_{k-3/2}^3 + \tfrac{3}{2} f_{k-2}^4 + \tfrac{7}{4} f_{k-5/2}^5 + \cdots$$

$$w^4\left(\frac{\mathrm{d}^4 f}{\mathrm{d}t^4}\right)_k = \qquad\qquad\qquad\qquad f_{k-2}^4 + 2 f_{k-5/2}^5 + \cdots$$

$$w^5\left(\frac{\mathrm{d}^5 f}{\mathrm{d}t^5}\right)_k = \qquad\qquad\qquad\qquad\qquad\qquad f_{k-5/2}^5 + \cdots.$$

$$\text{(VIII.34)}$$

Substituting this in Equation (VIII.33) we get, up to the differences of the fifth order,

$$\Delta_{k+1/2} = f_k + \tfrac{1}{2} f_{k-1/2}^1 + \tfrac{5}{12} f_{k-1}^2 + \tfrac{3}{8} f_{k-3/2}^3 + \tfrac{251}{720} f_{k-2}^4 + \tfrac{95}{288} f_{k-5/2}^5.$$

$$\text{(VIII.35)}$$

As all the differences on the right side are known, we in fact can compute $\Delta_{k+1/2}$ and then derive the next value of the function, x_{k+1}, by

$$x_{k+1} = x_k + \Delta_{k+1/2}$$

which solves the problem.

Notice that this procedure can be repeated. In fact we have $t_{k+1} = t_k + w$, so that we now can find $f_{k+1} = w \cdot F(t_{k+1}, x_{k+1})$ and then derive the next series of differences

$$f_{k+1/2}^1, \quad f_k^2, \quad f_{k-1/2}^3, \cdots,$$

thus adding a new ascending diagonal to our table. It is in this way that, step by step, the integration can be pushed forward as far as needed, or, alternatively, almost to the limit of the interval in which the differential equation is defined.

Let us add some remarks. The first one concerns the order of the highest difference to be retained. As to this point notice that the differences of the highest order we have retained in our integration tables are almost constant. If they were exactly constant, then the differences of the next highest order would be equal to zero. In view of the continuity of the function f, we shall ascribe the small variations of the values of the last differences retained to the influence of the rounding-off errors with which any value of f is necessarily affected. In our (main) integration Tables VIII.4 and VIII.5 these errors may have any size between $+0.00005$ and -0.00005. Moreover they build up in the differences of successively higher order and assume a leading role in the order where the differences alternate in sign. It is for this reason that the last significant differences will be the differences of the next lower order. Verify that in our tables the differences of an order higher than the last conserved effectively alternate in sign, so that one may say that they fluctuate around zero. The effect described bears some resemblance to the detection of a useful signal in the presence of noise. The last significant differences define the limit not to be exceeded, as by going further we only 'amplify the noise'.

Our second remark concerns the very principle on which the method is based. As to this point it must be noticed that what we effectively have done is an extrapolation. In fact, by using values at the end of a table we have computed a new one which is beyond the limit of this table. As extrapolation is always a risky undertaking it is reasonable to seek for a method permitting us to check and if necessary to correct each new value derived in this way. This can in fact be done by the following formula analogous to (VIII.35) and derived in Appendix II:

$$\Delta_{k+1/2} = f_{k+1} - \tfrac{1}{2} f_{k+1/2}^1 - \tfrac{1}{12} f_k^2 - \tfrac{1}{24} f_{k-1/2}^3$$
$$- \tfrac{19}{720} f_{k-1}^4 - \tfrac{3}{160} f_{k-3/2}^5 - \cdots . \qquad \text{(VIII.36)}$$

This formula is used as follows. First, as explained above, determine x_{k+1} and derive f_{k+1} as well as the whole set of the differences $f_{k+1/2}^1, f_k^2, \ldots$. Then by formula (VIII.36) compute a new value of the increment. If this value of the increment, which we shall denote by $(\Delta_{k+1/2})_{\text{corr}}$ coincides with the value found by extrapolation, then, obviously, x_{k+1} needs no correction. If this is not the case, then a new value of x_{k+1} must be computed by putting

$$x_{k+1} = x_k + (\Delta_{k+1/2})_{\text{corr}} . \qquad \text{(VIII.37)}$$

Use this value of x_{k+1} to derive $f_{k+1}, f_{k+1/2}^1, f_k^2, \ldots$ anew. It may happen that they coincide with the previous ones. In this case the cycle of iteration is closed. Should this not be the case, then $(\Delta_{k+1/2})_{\text{corr}}$ must be computed anew with the values of f_{k+1}, $f_{k+1/2}^1, \ldots$ just derived. This iterative process must be repeated until two successive iterations lead to the same value of the increment.

The method based on formula (VIII.35) is called the Adams' extrapolation method and the formula itself the predictor formula. The method based on the formula (VIII.36) is Adams' interpolation method, and the formula is called the corrector formula. Our main integration tables have been computed by making use of the predictor formula (VIII.35) which is simpler in application and which in our case yields sufficiently accurate results.

Our third remark is connected with the fact that the present (initial) position of the star is somewhere between the pericenter and the apocenter towards which it is moving. In order to determine the arc of the orbit between the pericenter and the apocenter we obviously have to integrate from the initial position forwards up to $R = R_{\text{max}}$, as well as backwards up to $R = R_{\text{min}}$.

Our final remark concerns some points which at first sight may seem only of theoretical interest but which, in fact, are of prime importance when effectively carrying out the integration of Equation (VIII.27). In this respect let us again recall that this equation is defined only in the interval $R_{min} \leqslant R \leqslant R_{max}$. Beyond these limits it is not valid, as the sign of the square root must be inverted. This means that values $R > R_{max}$ or $R < R_{min}$ computed according to our integration scheme are devoid of sense. When, in the course of its calculations, the reader arrives in the immediate vicinity of these limiting values, he will notice that, as expected, the quantity $f = w(dR/d\varphi)$ becomes smaller and smaller in absolute value. But he probably will be rather surprised by the fact that its value has become very sensitive to small variations in R. Moreover the hitherto smooth run of the differences $f^1, f^2, \dots$ will be disturbed (this is especially conspicuous in the highest order differences). At any other point such an effect, if it should occur, would almost inevitably indicate that a computation error has been made which must be corrected. When this has been done, the abrupt change in the differences will disappear. In the immediate proximity of the boundaries of the interval in which Equation (VIII.27) is defined, the effect described must be expected. In fact, we have by definition

$$f = w \frac{dR}{d\varphi}$$

so that, for a small variation δf in f we have

$$\delta f = w \frac{d}{d\varphi}\left(\frac{dR}{d\varphi}\right)\delta\varphi = w \frac{d}{dR}\left(\frac{dR}{d\varphi}\right) \cdot \frac{dR}{d\varphi}\delta\varphi = w \frac{d}{dR}\left(\frac{dR}{d\varphi}\right)\delta R.$$

Now it is easily proved that the expression for $(d/dR)(dR/d\varphi)$ contains a term in which the square root

$$\left[\frac{3.982R^2}{66.67 + R^2} - 0.0123R^2 - 1\right]^{1/2}$$

appears in the denominator. As the boundaries of the integration interval are identical to its zeros, the corresponding term necessarily becomes very large for values of R close to either of them. Therefore a small variation of R in fact must strongly affect f, and even more so the higher order differences $f^1, f^2, \dots$, making the computations at this point very uncertain.

Although very disappointing at first sight, this is not so serious as it may seem. In fact, it can be shown that, by adopting a smaller value for the integration step we can come closer to either of the boundary values of R, so that the values of φ which correspond to the pericenter and the apocenter respectively, can be determined with sufficient accuracy.

This remark is intended to warn the reader that a numerical integration does not consist only of the repetition of identical computations, but that it requires his full attention. When made by a fully automatic computer, it can lead to trustworthy results only if all the necessary checks have been incorporated in the program.

The reader is expressly advised to recalculate a fair number of values given in our (main) integration Tables VIII.4 and VIII.5, as numerical methods can never be understood by merely reading.

7. Technically speaking, the problem considered in the preceding section is that of continuing the numerical integration. When use is made of Adams' method, a different procedure must be sought to start the solution, i.e. to compute, starting from t_0, x_0, and f_0, an initial series of values of the argument, the function, of f_k as well as of its higher order differences, so that the formula (VIII.35) can be applied.

There are different methods to start an integration. In the present exercise we shall make use of the method proposed by A. N. Krylov, which is based on the following formulae (see, for the derivation, Appendix II)

$$\Delta_{1/2} = f_0 + \tfrac{1}{2} f^1_{1/2} - \tfrac{1}{12} f^2_1 + \tfrac{1}{24} f^3_{3/2}$$
$$\Delta_{3/2} = f_1 + \tfrac{1}{2} f^1_{1/2} + \tfrac{5}{12} f^2_1 - \tfrac{1}{24} f^3_{3/2} \qquad\qquad \text{(VIII.38)}$$
$$\Delta_{5/2} = f_2 + \tfrac{1}{2} f^1_{3/2} + \tfrac{5}{12} f^2_1 + \tfrac{3}{8} f^3_{3/2}$$

which are exact up to terms of the third order included. This means that the increments computed according to the formulae can be considered as exact if the differences of the third order are constant. Should they not be constant, then we must adopt a smaller value for the integration step so that this condition is satisfied.

Let us explain Krylov's method by effectively starting the integration backwards. Again, and for reasons already stated, we shall make the computations with a precision much higher than is really needed. We shall compute our starting table to five decimal places, the main tables to four places. After a few checks we find that a suitable value for the integration step is $w = -0.05$.

At the beginning we only know the initial values of the argument and of the function, which are

$$\varphi_0 = 0.00 \qquad R_0 = 10.00000.$$

In the present case the function on the right side of the differential equation is a function of R only. By substituting R_0 into Equation (VIII.27) and by performing the operations indicated, we get $f_0 = w (dR/d\varphi)_0 = -0.19947$. The first row of our starting table will therefore be

φ	R	f
0.00	10.00000	-0.19947

In order to find $R_1 = R_0 + \Delta_{1/2}$, we shall make use of the first of the formulae VIII.38 reduced to its first term, i.e. put $\Delta_{1/2} = f_0$. This leads to the following table in which the known quantities are printed in bold type

φ	R	Δ	f	f^1
0.00	**10.00000**		**−0.19947**	
		-0.19947		-202
-0.05	9.80053		-0.20149	

Here $f_1 = w (dR/d\varphi)_1$, and $f^1_{1/2} = f_1 - f_0$. In order to save space we have written in the column f^1 simply -202 for -0.00202.

Now the values derived can only be considered as provisional ones, and are subject to improvement by iteration. For that purpose we shall compute R_1 anew, but now take the first two terms in the formula for $\Delta_{1/2}$. We have

$$\Delta_{1/2} = f_0 + \tfrac{1}{2} f_{1/2}^1 = -0.19947 - 0.00101 = -0.20048.$$

This gives the following table where again the quantities considered as given have been printed in bold type:

0.00	**10.00000**		**− 0.19947**	
		− 0.20048		− 202
− 0.05	9.79952		− 0.20150	

The new value for f^1 practically coincides with the previous one, as by rounding off $\tfrac{1}{2}(-0.00203)$ we again get -0.00101. Therefore further iterations are impossible and we can make the next step by deriving a first approximate value for R_2. For that purpose we shall make use of the second of the Equations (VIII.38) reduced to its first two terms. This gives

$$\Delta_{3/2} = -0.20150 - 0.00101 = -0.20251$$

and

φ	R	Δ	f	f^1	f^2
				− 203	
− 0.05	**9.79952**		**− 0.20150**		+ 162
		− 0.20251		− 41	
− 0.10	9.59701		− 0.20191		

The value of f_1^2 which we can now determine is quite large. Therefore we are compelled to correct all the values computed so far. Obviously we shall make use of the same formulae as before, but we can now take the first three terms in each of them. In other words we shall put

$$\Delta_{1/2} = f_0 + \tfrac{1}{2} f_{1/2}^1 - \tfrac{1}{12} f_1^2$$
$$\Delta_{3/2} = f_1 + \tfrac{1}{2} f_{1/2}^1 + \tfrac{5}{12} f_1^2 .$$

The starting table then becomes

0.00	**10.00000**		**− 0.19947**		
		− 0.20062		− 203	
					+ 162
− 0.05	9.79938		− 0.20150	— — —	
					+ 161
				− 41	
		− 0.20184		— — —	
					− 42
− 0.10	9.59754		− 0.20192.		

The difference between the new values of $f_{3/2}^1$ and f_1^2 (given below the broken line) and the previous ones is so small that it will not affect other values in this table. We therefore can proceed a step further and compute $\varDelta_{5/2}$ by

$$\varDelta_{5/2} = f_2 + \tfrac{1}{2} f_{3/2}^1 + \tfrac{5}{12} f_1^2 = -0.20192 + \tfrac{1}{2} \times (-0.00042)$$
$$+ \tfrac{5}{12} \times 0.00161 = -0.20146$$

which gives for the next following value of the function

$$R_3 = R_2 + \varDelta_{5/2} = 9.59754 - 0.20146 = 9.39608 .$$

We now can extend our starting table by this new value of R, find f_3, and derive a set of differences:

$$f_3 = w(dR/d\varphi)_3 = -0.20088 ; \qquad f_{5/2}^1 = f_3 - f_2 = +0.00104 ;$$
$$f_2^2 = f_{5/2}^1 - f_{3/2}^1 = +0.00146 ; \qquad f_{3/2}^3 = f_2^2 - f_1^2 = -0.00015 .$$

Obviously the quantities determined so far must be corrected once again taking into account the third order difference, but the corrections will be rather small. Once this has been done we can extend our table by a couple of steps, making use of the formula (VIII.35). In this way we get Table VIII.3.

TABLE VIII.3

φ	R	$\varDelta$	f	f^1	f^2	f^3
0.00	10.00000		−0.19947			
		−0.20063		−203		
−0.05	9.79937		−0.20150		+161	
		−0.20183		−42		−15
−0.10	9.59754		−0.20192		+146	
		−0.20152		+104		−16
−0.15	9.39602		−0.20088		+130	
		−0.19981		+234		−15
−0.20	9.19621		−0.19854		+115	
		−0.19689		+349		−17
−0.25	8.99932		−0.19505		+98	
		−0.19288		+447		
−0.30	8.80644		−0.19058			

As seen, the differences of the third order are quite constant. Therefore this table can be considered as the definitive starting table. By rounding off the last values to four decimal places we get the first series of values for our main integration Table VIII.5. It is not difficult to see that from Table VII.3 we also can derive the starting diagonal for forward integration given at the top of the main forward integration Table VIII.4. For this purpose consider provisionally $\varphi = -0.3$ and $R(-0.3)$ as the initial values and move forward ($w > 0$). Change the signs accordingly.

In general it will be found convenient to arrange the integration so that the differences of the fourth order can be neglected. If in the run of the computation they should become significant, then it is advisable to adopt a smaller integration step. However this implies a new starting procedure. The same must be done (but carefully) if, starting from some value not too close to the boundaries one will come closer to the boundaries than was possible with the value of the step previously adopted.

It is proposed that the reader should integrate the differential Equation (VIII.27) adopting for the step $w=0.1$ and computing to three decimal places. He should make use of the predictor formula (VIII.35).

Give your solution in the form of a table where the galactocentric distances R, in kpc, will be tabulated as a function of the azimuthal angle φ, expressed in degrees. Retain only the figures common to your and our solutions.

Make a drawing of the arc of orbit of 61 Cygni according to your solution, and convince yourself that the orbit has the form of a rosette. It can also be described as an ellipsoidal curve exhibiting a strong apsidal motion. Give an estimate of the angle between the direction towards the pericenter and the next following apocenter (for pure elliptical motion this angle is equal to 180 deg).

Compute the time ΔT the star needs to describe the arc of its orbit between the points $\varphi = -1.6$ rad., $R = R(-1.6)$ and $\varphi = 0.6$ rad, $R = R(0.6)$, which roughly corresponds to the time between two consecutive passages through the pericenter and the apocenter. For that purpose evaluate the integral (compare Equation (VIII.3))

$$h \cdot \Delta T = \int_{-1.6}^{+0.6} R^2 \, d\varphi \qquad (VIII.41)$$

by making use of some of the usual formulae (e.g. Simpson's). Notice that, as the unit of distance and the unit of velocity are respectively one kpc and one km/s the time unit is fixed and in fact equal to

$$1 \text{ kpc}/(1 \text{ km/s}) = 0.978 \times 10^9 \text{ yr}.$$

Problem VIII.1. Derivation of the General Differential Equation of the Motion of a Star in the Galaxy

Derive the Equations (VIII.1).

Solution

Denote by $q_j\,(j=1,2,3)$ the generalised coordinates of a star. Lagrange's equations of motion will be

$$\frac{d}{dt}\left(\frac{\partial T}{\partial \dot{q}_j}\right) - \frac{\partial T}{\partial q_j} = Q_j, \quad j = 1, 2, 3, \tag{P.VIII.1}$$

where T denotes the kinetic energy of the star and where Q_j is the component of the generalised force corresponding to the generalised coordinate q_j. As it will be seen one can assume without loss of generality that the mass of the star is equal to unity. The kinetic energy will then be given by

$$T = \tfrac{1}{2}(\dot{R}^2 + R^2\dot{\varphi}^2 + \dot{z}^2). \tag{P.VIII.2}$$

For the derivatives on the left side of the Equations (P. VIII.1) we therefore get

$$\frac{\partial T}{\partial \dot{R}} = \dot{R} \qquad \frac{\partial T}{\partial \dot{\varphi}} = R^2\dot{\varphi} \qquad \frac{\partial T}{\partial \dot{z}} = \dot{z}$$

$$\frac{\partial T}{\partial R} = R\dot{\varphi}^2 \qquad \frac{\partial T}{\partial \varphi} = 0 \qquad \frac{\partial T}{\partial z} = 0. \tag{P.VIII.3}$$

The differential equations of motion in cylindrical coordinates will then be

$$\ddot{R} - R\dot{\varphi}^2 = F_R$$

$$\frac{d}{dt}(R^2\dot{\varphi}) = F_\varphi \tag{P.VIII.4}$$

$$\ddot{z} = F_z$$

where F_R, F_φ, and F_z are the components of the generalised force per unit mass.

Now we have assumed that the force acting upon the star is of gravitational character. This means that it can be derived from a potential Φ. In general this potential will be a function of all the three coordinates R, φ, and z. In other words, in the general case we shall have

$$F_R = \frac{\partial \Phi}{\partial R}, \qquad F_\varphi = \frac{\partial \Phi}{\partial \varphi}, \qquad F_z = \frac{\partial \Phi}{\partial z}. \tag{P.VIII.5}$$

In the present case, however, the potential Φ will not depend on φ. In fact take for R and z two fixed, but otherwise arbitrary values. Then any variation of φ cannot affect Φ as, due to the rotational symmetry of our stellar system the same properties will be found in all points of the circle $R=$const, $z=$const, $0\leqslant\varphi\leqslant 2\pi$.

Problem VIII.2. Differential Equation of the Motion of a Star in the Galactic Plane

Prove the Equations (VIII.2)

Solution

Consider the third of the differential equations of the motion (VIII.1). Obviously F_z is the component of the force along the z-axis. As according to one of our basic assumptions the distribution of the attracting masses is symmetric with respect to the galactic plane (i.e. to the plane $z=0$), this component must vanish in the plane itself. Moreover it is not difficult to see that in our stellar system this force will always be directed towards the galactic plane.

Take now a star which at some initial moment $t=t_0$ is in the galactic plane so that its z coordinate is equal to zero: $z(t_0)=0$. According to the third of the Equations (VIII.1) we will have $\dot{z}=C$, where C is an integration constant. Now $\dot{z}$ is the component of the star's velocity along the z-axis or perpendicular to the galactic plane. If now not only $z(t_0)=0$ but also $\dot{z}(t_0)=0$ then necessarily $C=0$, so that z will remain equal to zero. This means that the star will describe an orbit in the galactic plane. In this case the system of equations of motion in fact reduces to (VIII.2).

Problem VIII.3.   The Energy Integral

Find the second integral (VIII.4) and show that it is identical to the energy integral.

Solution

The second integral can be found as follows. In the first of the equations (VIII.2) substitute for $\dot{\varphi}$ its value from (VIII.3). This gives

$$\ddot{R} = (h^2/R^3) + F_R. \tag{P.VIII.6}$$

Now in the case of plane motion $z \equiv 0$ so that Φ will in fact be a function of R only, $\Phi = \Phi(R)$. Therefore

$$F_R = d\Phi/dR. \tag{P.VIII.7}$$

Now multiply (P. VIII.6) by $dR = \dot{R}\, dt$ and make use of (P. VIII.7). Then

$$\dot{R}\ddot{R}\, dt = (h^2/R^3)\, dR + F_R\, dR = (h^2/R^3)\, dR + d\Phi, \tag{P.VIII.8}$$

an expression which can immediately be integrated giving

$$\tfrac{1}{2}\dot{R}^2 = -\tfrac{1}{2}(h^2/R^2) + \Phi + H, \qquad \Phi = \Phi(R) = \int F_R\, dR + C. \tag{P.VIII.9}$$

The quantity H is a new integration constant. Like h it can be determined from the initial conditions.

It can easily be shown that (P. VIII.9) represents the energy integral. In fact, writing (P. VIII.9) as

$$\tfrac{1}{2}(\dot{R}^2 + h^2/R^2) = \Phi(R) + H \tag{P.VIII.10}$$

and substituting for h its value given by (VIII.3) we get

$$\tfrac{1}{2}(\dot{R}^2 + R^2\dot{\varphi}^2) - \Phi(R) = H. \tag{P.VIII.11}$$

According to (P. VIII.2) the first term on the left side of (P. VIII.11) is the kinetic energy of the star, whereas $-\Phi(R)$ is its potential energy. Therefore this equation in fact states that the sum of the kinetic and of the potential energy is constant.

References

[1] Eggen, O. J., Lynden-Bell, D., and Sandage, A. R.: 1962, *Astrophys. J.* **136**, 748.
[2] Strömgren, B.: 1963, *Quart. J. Roy. Astron. Soc.* **4**, 8.
[3] Kelsall, T. and Strömgren, B.: 1966, *Vistas Astron.* **8**, 159.
[4] Torgard, I.: 1956, *Medd. Lunds Astron. Obs.*, Ser. II, No. 133.
[5] Ollongren, A.: 1965, in A. Blaauw and M. Schmidt (eds.), *Galactic Structure*, University of Chicago Press, Chicago.
[6] Contopoulos, G.: 1956, *Astrophys. J.* **124**, 643.
[7] Trumpler, R. and Weaver, H. F.: 1953, *Statistical Astronomy*, Ch. 6.1, University of California Press, Berkeley. Reprint by Dover Publications, New York, 1962.

[8] Ogorodnikov, K. F.: 1965, *Dynamics of Stellar Systems*, Pergamon Press, Oxford.

[9] Contopoulos, G. and Strömgren, B.: 1965, *Tables of Plane Galactic Orbits*, Goddard Space Flight Center, NASA, New York.

[10] Parenago, P. P.: 1950, *Astron. Zh.* **27**, 150. German translation in *Abhandl. Sowjetischen Astron.*, Series II, Verlag Kultur und Fortschritt, D.D.R., Berlin 1951, p. 57.

[11] Delhaye, J.: 1965, in A. Blaauw and M. Schmidt (eds.), *Galactic Structure*, University of Chicago Press, Chicago, p. 61.

[12] Mihalas, D., with the collaboration of McRae Routly, P.: 1968, *Galactic Astronomy*, Freeman and Co., San Francisco.

[13] Parenago, P. P.: 1950, *Astron. Zh.* **27**, 329. German translation in *Abhandl. Sowjetischen Astron.*, Series II, Verlag Kultur und Fortschritt, D.D.R., Berlin, 1951, p. 81.

[14] Eggen, O. J.: 1962, *Roy. Obs. Bull.* No. 51, HM Stationery Office, London.

[15] Chandrasekhar, S.: 1960, *Principles of Stellar Dynamics*, Dover Publ., New York.

TABLE VIII.4[a]

φ	R	Δ	f	f^1	f^2	f^3
						-2
					-16	
				-20		-2
0.00	10.0000		$+0.1995$		-18	
		$+0.1978$		-38		0
$+0.05$	10.1978		$+0.1957$		-18	
		$+0.1930$		-56		(-3)
$+0.10$	10.3908		$+0.1901$		-21	
		$+0.1866$		-77		$(+2)$
$+0.15$	10.5774		$+0.1824$		-19	
		$+0.1777$		-96		(-1)
$+0.20$	10.7551		$+0.1728$		-20	
		$+0.1672$		-116		(-1)
$+0.25$	10.9223		$+0.1612$		-21	
		$+0.1546$		-137		$(+1)$
$+0.30$	11.0769		$+0.1475$		-20	
		$+0.1398$		-157		$+3$
$+0.35$	11.2167		$+0.1318$		-17	
		$+0.1231$		-174		0
$+0.40$	11.3398		$+0.1144$		-17	
		$+0.1051$		-191		$+3$
$+0.45$	11.4449		$+0.0953$		-14	
		$+0.0850$		-205		$+3$
$+0.50$	11.5299		$+0.0748$		-11	
		$+0.0641$		-216		$+2$
$+0.55$	11.5940		$+0.0532$		-9	
		$+0.0421$		-225		
$+0.60$	11.6361		$+0.0307$			
		$+0.0192$				
$+0.65$	11.6553		$(+0.0079)$			

[a] Values in brackets have not been taken into account.

TABLE VIII.5[a]

φ	R	Δ	f	f^1	f^2	f^3
						-2
					$+10$	
				$+45$		-2
-0.30	8.8064		-0.1906		$+8$	
		-0.1880		$+53$		-1
-0.35	8.6184		-0.1853		$+7$	
		-0.1824		$+60$		-1
-0.40	8.4360		-0.1793		$+6$	
		-0.1760		$+66$		-1
-0.45	8.2601		-0.1727		$+5$	
		-0.1692		$+71$		-2

[a] Values in brackets have not been taken into account. For $0 \geqslant \varphi \geqslant -0.25$ see Table VIII.3, p. 124.

Table VIII.5 (continued)

φ	R	Δ	f	f^1	f^2	f^3
−0.50	8.0908		−0.1656		+3	
		−0.1619		+74		−1
−0.55	7.9289		−0.1582		+2	
		−0.1545		+76		(+1)
−0.60	7.7744		−0.1506		+3	
		−0.1468		+79		(−2)
−0.65	7.6276		−0.1427		+1	
		−0.1386		+80		(−1)
−0.70	7.4890		−0.1347		0	
		−0.1307		+80		(0)
−0.75	7.3583		−0.1267		0	
		−0.1227		+80		
−0.80	7.2356		−0.1187		(+1)	
		−0.1147		+81		
−0.85	7.1209		−0.1106		(−2)	
		−0.1065		+79		
−0.90	7.0144		−0.1027		(+1)	
		−0.0988		+80		
−0.95	6.9156		−0.0947		−2	
		−0.0907		+78		
−1.00	6.8249		−0.0869		−1	
		−0.0831		+77		
−1.05	6.7418		−0.0792		0	
		−0.0754		+77		
−1.10	6.6664		−0.0715		−2	
		−0.0676		+75		
−1.15	6.5988		−0.0640		−1	
		−0.0603		+74		
−1.20	6.5385		−0.0566		0	
		−0.0529		+74		
−1.25	6.4856		−0.0492		−2	
		−0.0455		+72		
−1.30	6.4401		−0.0420		0	
		−0.0385		+72		
−1.35	6.4016		−0.0348		−1	
		−0.0312		+71		
−1.40	6.3704		−0.0277		−1	
		−0.0242		+71		
−1.45	6.3462		−0.0206		−1	
		−0.0171		+70		
−1.50	6.3291		−0.0136			
		−0.0101				
−1.55	6.3190		(−0.0065)			

ᵃ Values in brackets have not been taken into account.

APPENDIX I

DERIVATION OF THE NORMAL EQUATIONS (III.15) BY
THE PRINCIPLE OF LEAST SQUARES

1. Suppose that we effectively have computed the coefficients a_i, b_i, and c_i, for all the cluster members or, alternatively, only for a number of them, selected according to some criterion. Let us write down the corresponding Equations (III.13). Then we shall get the following system of equations:

$$a_i x + b_i y = c_i, \quad i = 1, 2, ..., n. \tag{A.I.1}$$

The Equations (A.I.1) may be called the equations of condition of our problem.

Now, if the values of the proper motion components were exactly known and if the velocity vectors were strictly parallel, then, by simple algebra, we could determine the quantities x and y from any two of the equations of the system (A.I.1), the remaining $n-2$ equations being rigorously satisfied by these values of the unknowns. Unfortunately this is not the case. As seen from Table III.1 the values of the proper motions are affected by observational errors which, in general, are rather small but not negligible. Therefore, even if the possible spread in the directions of the velocities of the Hyades stars can be neglected (which seems permissible), each pair of the Equations (A.I.1) necessarily must lead to a different set of values for the unknowns, and none of the remaining $n-2$ equations will strictly be satisfied by these values for x and y. In other words, instead of being exactly equal to zero the remaining $n-2$ expressions

$$a_i x + b_i y - c_i$$

will have a finite value.

Obviously there is no reason to give preference to any of all the possible solutions which can be found in this way. Moreover, in view of the fact that the proper motions are affected by errors, we even cannot consider as especially reliable a solution based on two stars only. However we shall certainly come closer to the truth if in some way we could derive x and y by taking into account the whole set of Equations (A.I.1). In fact we can expect that in this case the effect of random errors in the values of the proper motions (as well as that of possible random deviation of the velocities from strict parallelism) will be largely eliminated.

Now we have seen that it is impossible to find a set of values for x and y satisfying all the equations of condition rigorously. Let us, then, try to determine x and y so that they all will be satisfied as nearly as possible. But how shall we decide that a given set of values for x and y meets this requirement better than all the others? Usually one adopts Legendre's Principle of least squares according to which one shall take, as the best, the values for x and y which render the sum of the squares

$$Q = \sum_{i=1}^{n} (a_i x + b_i y - c_i)^2 \qquad (A.I.2)$$

a minimum.

Notice that the value of Q depends on the values adopted for x and y so that we can write $Q = Q(x, y)$; that Q must always be positive, vanishing only if all the equations of condition can be satisfied rigorously.

It is now a simple matter to determine the 'best' value for x and y, as they can be found from the well-known condition for an extreme value of Q, which necessarily must be a minimum. We have

$$\frac{\partial Q}{\partial x} = \frac{\partial}{\partial x} \sum_{i=1}^{n} (a_i x + b_i y - c_i)^2 = 0, \quad \frac{\partial Q}{\partial y} = \frac{\partial}{\partial y} \sum_{i=1}^{n} (a_i x + b_i y - c_i)^2 = 0$$

$$(A.I.3)$$

or by performing the operations indicated

$$\left(\sum_{i=1}^{n} a_i^2 \right) x + \left(\sum_{i=1}^{n} a_i b_i \right) y = \sum_{i=1}^{n} a_i c_i$$

$$\left(\sum_{i=1}^{n} b_i a_i \right) x + \left(\sum_{i=1}^{n} b_i^2 \right) y = \sum_{i=1}^{n} b_i c_i. \qquad (A.I.4)$$

The Equations (A.I.4) are the so-called normal equations, which solve our problem. Notice that the number of normal equations must necessarily be equal to the number of unknowns.

Instead of the cumbersome explicit form for the coefficients, one usually takes the following ones, due to Gauss, who independently introduced the principle of least squares:

$$\sum_{i=1}^{n} a_i^2 = [aa], \qquad \sum_{i=1}^{n} a_i b_i = [ab], \qquad \sum_{i=1}^{n} a_i c_i = [ac]$$

$$\sum_{i=1}^{n} b_i a_i = [ba], \qquad \sum_{i=1}^{n} b_i^2 = [bb], \qquad \sum_{i=1}^{n} b_i c_i = [bc].$$

2. The determination of the coefficients of the normal equations involves a rather large amount of computational work. Therefore it will be necessary to check the values obtained. This can be done as follows.

Compute not only the quantities a_i, b_i, and c_i, but also their sums

$$a_i + b_i + c_i = s_i. \qquad (A.I.5)$$

In our example they are all given in Table III.3, p. 47. In order to check the quantities s_i notice that by (A.I.5) the sum $\sum s_i$ of the numbers standing in the last column of the Table III.3 must be equal to

$$\sum_{i=1}^{n} a_i + \sum_{i=1}^{n} b_i + \sum_{i=1}^{n} c_i$$

that is equal to the result obtained by adding the sums of the three preceding columns.

As to the coefficients of the first normal equation we have

$$a_i^2 + a_i b_i + a_i c_i = a_i s_i$$

and

$$\sum_{i=1}^n a_i^2 + \sum_{i=1}^n a_i b_i + \sum_{i=1}^n a_i c_i = \sum_{i=1}^n a_i s_i. \qquad (A.I.6)$$

Therefore the values of the coefficients of the first normal equation, which are given in the last row of Table III.4 can be checked in the same way. In fact, by (A.I.6) their sum must be equal to the result obtained by adding the numbers in the last column of Table III.4. The same applies to Table III.5 which contains the coefficients of the second normal equation.

Up to this stage one must conserve all the digits obtained in the computations as otherwise it is not possible to decide whether or not an eventual difference between two check sums may be imputed to an error. However once the coefficients have been checked their values can be rounded off to the significant number of digits.

Let us finally add that the problem considered here in connection with the determination of the vertex of the Hyades cluster is a very common one in astronomy. In fact, it arises always when one has to determine some unknown quantities, say, x, y, z, ..., which cannot be directly measured, but which are known to be related to some other, measurable quantities a, b, c, ..., n, by linear relations of the form

$$ax + by + cz + ... = n.$$

The reader will find another example at the end of Exercise III. A similar case is the determination of Oort's first constant of galactic rotation, as well as of the components of the solar motion, discussed in Exercise VII, Section 6 (see Equation (VII.32)). All such problems are usually solved by the method outlined above.

It must be stressed, however, that from a mathematical point of view there is an essential difference between the determination of the quantities x and y by Charlier's method on the one side, and, on the other, the determination of the corrections ΔA and ΔD, briefly discussed at the end of Exercise III. In fact in the first case all the coefficients which appear in the equations of condition (III.13) are affected by errors of the same order of magnitude, as they all are computed from the observed proper motions (by the Equations (III.14)). In the second case, however, these errors affect only the quantities $\Delta\theta$ on the left side of the equations of condition (III.22). It is only in cases like this last one that the method of least squares yields correct values of the unknowns.

For a discussion of this question the more interested reader is referred to the monograph by Trumpler and Weaver (see references on p. XI), Chapter 1.7, in particular §1.74. Also, he is expressly advised to consult, for a full account of the method of least squares, some of the standard text-books. An especially useful reference is B. M. Schchigolev's excellent *Mathematical Treatment of Observations* (Iliffe books, London, 1965, translated from Russian).

PROOF OF THE FORMULAE VIII.36, VIII.38, AND VIII.40

1. The relations VIII.36 follow from Newton's formula for backward interpolation. For the convenience of these readers who are not familiar with interpolation, we shall first give a short derivation of both Newton's interpolation formulae.

Let $f = f(t)$ be a function of the argument t, given in tabular form for equidistant values of t. Put $t_{j+1} - t_j = w$, and $f_j = f(t_j); j = 0, 1, 2, \ldots$. Compute the successive order differences of f as explained in Exercise VIII, Section 6. This leads to Table A.II.1

TABLE A.II.1

t	f	f^1	f^2	f^3	$\ldots$
t_0	f_0				
		$f^1_{1/2}$			
t_1	f_1		f^2_1		
		$f^1_{3/2}$		$f^3_{3/2}$	
t_2	f_2		f^2_2		$\ldots$
		$f^1_{5/2}$		$\ldots$	
t_3	f_3		$\ldots$		
		$\ldots$			
$\ldots$	$\ldots$				

Now we have

$$f_0 = f_0$$
$$f_1 = f_0 + f^1_{1/2}$$
$$f_2 = f_1 + f^1_{3/2} = (f_0 + f^1_{1/2}) + (f^1_{1/2} + f^2_1) = f_0 + 2f^1_{1/2} + f^2_1.$$

In the same way we find

$$f_3 = f_0 + 3f^1_{1/2} + 3f^2_1 + f^3_{3/2}, \ldots \quad \text{and so on}.$$

The coefficients appearing in the expressions for the successive values of the function are obviously the binomial coefficients, so that in general we will have

$$f_n = f(t_0 + nw) = f_0 + \binom{n}{1} f^1_{1/2} + \binom{n}{2} f^2_1 + \binom{n}{3} f^3_{3/2} + \cdots. \quad \text{(A.II.1)}$$

For a value of the argument t which lies "between the lines" of the table we shall put $t = t_0 + z \cdot w$, where $z = (t - t_0)/w$ will not, in general, be a whole number. Formula

(A.II.1) then becomes

$$f(t_0 + zw) = f_0 + \frac{z}{1!}f^1_{1/2} + \frac{z(z-1)}{2!}f^2_1 + \frac{z(z-1)(z-2)}{3!}f^3_{3/2} + \cdots.$$

(A.II.2)

If instead from t_0 we start from some arbitrary value of the argument, say t_k, we obviously will have

$$f_{k+n} = f(t_k + nw) = f_k + \binom{n}{1}f^1_{k+1/2} + \binom{n}{2}f^2_{k+1} + \binom{n}{3}f^3_{k+3/2} + \cdots$$

(A.II.3)

and

$$f(t_k + zw) = f_k + \frac{z}{1!}f^1_{k+1/2} + \frac{z(z-1)}{2!}f^2_{k+1}$$
$$+ \frac{z(z-1)(z-2)}{3!}f^3_{k+3/2} + \cdots.$$

(A.II.4)

This is Newton's formula for forward interpolation. Notice that the differences which it contains are disposed along a descending diagonal in the Table A.II.1, whereas in Exercise VIII, Section 6, we have had differences lying on an ascending diagonal. The formula we need, and which is the so-called Newton's formula for backward interpolation, can be derived from (A.II.3) as follows.

Consider the Table A.II.2 where both the descending as well as the ascending dia-

TABLE A.II.2

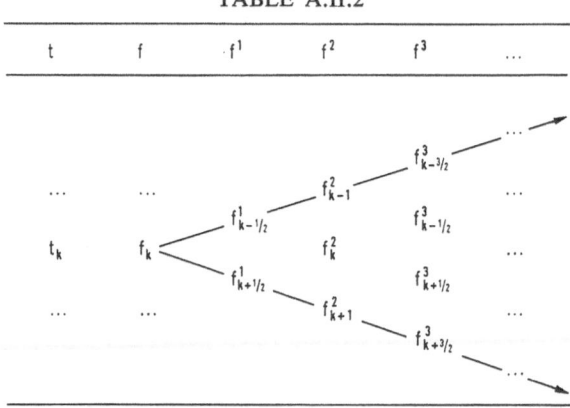

gonal have been drawn. All we have to do is to express the differences on the descending diagonal by those on the ascending one. In order to achieve this let us first suppose that the first-order differences are constant. This necessarily means that the differences of the second and of all the higher orders are equal to zero. Then formula (A.II.3) gives

$$f_{k+n} = f_k + \binom{n}{1}f^1_{k+1/2} = f_k + \frac{n}{1!}f^1_{k-1/2},$$

(A.II.5)

as, according to our assumption $f^1_{k+1/2} = f^1_{k-1/2} = $ constant.

Consider now the case when the second-order differences are constant. Then (A.II.3) gives

$$f_{k+n} = f_k + \binom{n}{1} f^1_{k+1/2} + \binom{n}{2} f^2_{k+1}.$$

(A.II.6)

Now according to our assumption $f^2_{k+1} = f^2_k = f^2_{k-1}$. Moreover we will have

$$f^1_{k+1/2} = f^1_{k-1/2} + f^2_k = f^1_{k-1/2} + f^2_{k-1}.$$

(A.II.7)

Notice that this last result may also be obtained by making use of the formula (A.II.5) just derived. In fact the quantities f^2 are the first-order differences of the f^1 ones, and, according to our assumption they are constant. For n we obviously shall take $n = (k + \frac{1}{2}) - (k - \frac{1}{2}) = 1$.

By substituting (A.II.7) into (A.II.6) we get

$$f_{k+n} = f_k + \binom{n}{1} f^1_{k-1/2} + \left[\binom{n}{1} + \binom{n}{2} \right] f^2_{k-1}$$

$$= f_k + \frac{n}{1!} f^1_{k-1/2} + \frac{n(n+1)}{2!} f^2_{k-1}.$$

(A.II.8)

Consider finally the case when the third order differences are constant. Then (A.II.3) reduces to

$$f_{k+n} = f_k + \binom{n}{1} f^1_{k+1/2} + \binom{n}{2} f^2_{k+1} + \binom{n}{3} f^3_{k+3/2}.$$

(A.II.9)

According to this new assumption we shall have $f^3_{k+3/2} = f^3_{k-3/2}$. For the second-order differences we get from (A.II.5), with $n = (k+1) - (k-1) = 2$

$$f^2_{k+1} = f^2_{k-1} + 2f^3_{k-3/2}.$$

(A.II.10)

We notice further that the second-order differences of the quantities f^1, i.e. the quantities f^3 are constant, so that use can be made of the formula (A.II.8). For n we shall have $n = (k + \frac{1}{2}) - (k - \frac{1}{2}) = 1$. Then (A.II.8) gives

$$f^1_{k+1/2} = f^1_{k-1/2} + f^2_{k-1} + f^3_{k-3/2}.$$

(A.II.11)

By substituting the values of $f^1_{k+1/2}, f^2_{k+1}$, and $f^3_{k+3/2}$, into (A.II.9) we get

$$f_{k+n} = f_k + \binom{n}{1} f^1_{k-1/2} + \left[\binom{n}{1} + \binom{n}{2} \right] f^2_{k-1}$$

$$+ \left[\binom{n}{1} + 2 \binom{n}{2} + \binom{n}{3} \right] f^3_{k-3/2}$$

which after some transformations gives

$$f_{k+n} = f_k + \frac{n}{1!} f^1_{k-1/2} + \frac{n(n+1)}{2!} f^2_{k-1} + \frac{n(n+1)(n+2)}{3!} f^3_{k-3/2}.$$

(A.II.12)

By comparing (A.II.12), (A.II.8), and (A.II.5), it is immediately seen that, in the general case, f_{n+k} will be given by

$$f_{k+n} = f_k + \frac{n}{1!} f^1_{k-1/2} + \frac{n(n+1)}{2!} f^2_{k-1} + \frac{n(n+1)(n+2)}{3!} f^3_{k-3/2} + \cdots .$$

$$\text{(A.II.13)}$$

For values of the argument which lie between the tabulated ones we shall again put $z = (t - t_k)/w$, which gives

$$f(t_k + zw) = f_k + \frac{z}{1!} f^1_{k-1/2} + \frac{z(z+1)}{2!} f^2_{k-1}$$

$$+ \frac{z(z+1)(z+2)}{3!} f^3_{k-3/2} + \cdots . \qquad \text{(A.II.14)}$$

This is Newton's formula for backward interpolation.
Now the Taylor expansion of $f(t_k + zw)$ is

$$f(t_k + zw) = f_k + \frac{zw}{1!} \left(\frac{df}{dt}\right)_k + \frac{z^2 w^2}{2!} \left(\frac{d^2 f}{dt^2}\right)_k + \frac{z^3 w^3}{3!} \left(\frac{d^3 f}{dt^3}\right)_k + \cdots .$$

On the other hand (A.II.14) can be written as

$$f(t_k + zw) = f_k + \frac{z}{1!} f^1_{k-1/2} + \frac{z^2 + z}{2!} f^2_{k-1} + \frac{z^3 + 3z^2 + 2z}{3!} f^3_{k-3/2}$$

$$+ \frac{z^4 + 6z^3 + 11z^2 + 6z}{4!} f^4_{k-2}$$

$$+ \frac{z^5 + 10z^4 + 35z^3 + 50z^2 + 24z}{5!} f^5_{k-5/2} + \cdots$$

or again as

$$f(t_k + zw)$$

$$= f_k + \frac{z}{1!}(f^1_{k-1/2} + \tfrac{1}{2} f^2_{k-1} + \tfrac{1}{3} f^3_{k-3/2} + \tfrac{1}{4} f^4_{k-2} + \tfrac{1}{5} f^5_{k-5/2} + \cdots)$$

$$+ \frac{z^2}{2!}(f^2_{k-1} + f^3_{k-3/2} + \tfrac{11}{12} f^4_{k-2} + \tfrac{5}{6} f^5_{k-5/2} + \cdots)$$

$$+ \frac{z^3}{3!}(f^3_{k-3/2} + \tfrac{3}{2} f^4_{k-2} + \tfrac{7}{4} f^5_{k-5/2} + \cdots)$$

$$+ \frac{z^4}{4!}(f^4_{k-2} + 2 f^5_{k-5/2} + \cdots)$$

$$+ \frac{z^5}{5!}(f^5_{k-5/2} + \cdots) + \cdots .$$

By comparing this last expression with the Taylor expansion we immediately get the formulae (VIII.36).

2. With the same notations as in Exercise VIII, Section 6, we have, by expanding $x(t)$ around $t=t_{k+1}$

$$x(t_k) = x(t_{k+1}) - \frac{w}{1!}\left(\frac{dx}{dt}\right)_{k+1} + \frac{w^2}{2!}\left(\frac{d^2x}{dt^2}\right)_{k+1} - + \cdots$$

or by virtue of (VIII.30) and (VIII.31)

$$\Delta_{k+1/2} = f_{k+1} - \frac{w}{2!}\left(\frac{df}{dt}\right)_{k+1} + \frac{w^2}{3!}\left(\frac{d^2f}{dt^2}\right)_{k+1} - + \cdots$$

so that now we shall have

$$
\begin{aligned}
\Delta_{k+1/2}
&= f_{k+1} - \tfrac{1}{2}(f^1_{k+1/2} + \tfrac{1}{2} f^2_k + \tfrac{1}{3} f^3_{k-1/2} + \tfrac{1}{4} f^4_{k-1} + \tfrac{1}{5} f^5_{k-3/2} + \cdots) \\
&\quad + \tfrac{1}{6}(f^2_k + f^3_{k-1/2} + \tfrac{11}{12} f^4_{k-1} + \tfrac{5}{6} f^5_{k-3/2} + \cdots) \\
&\quad - \tfrac{1}{24}(f^3_{k-1/2} + \tfrac{3}{2} f^4_{k-1} + \tfrac{7}{4} f^5_{k-3/2} + \cdots) \\
&\quad + \tfrac{1}{120}(f^4_{k-1} + 2 f^5_{k-3/2} + \cdots) \\
&\quad - \tfrac{1}{720}(f^5_{k-3/2} + \cdots) + - \cdots,
\end{aligned}
$$

which after some transformations leads to the corrector formula (VIII.36).

3. Let us finally derive Krylov's formulae for starting the integration. For this purpose consider the Table A.II.3 and suppose that the differences of the third order are con-

TABLE A.II.3

stant. In this case the usual formula for the increment, viz.

$$\Delta_{k+1/2} = f_k + \tfrac{1}{2} f^1_{k-1/2} + \tfrac{5}{12} f^2_{k-1} + \tfrac{3}{8} f^3_{k-3/2}$$

may be written as follows

$$\Delta_{k+1/2} = f_k + \tfrac{1}{2} f^1_{k-1/2} + \tfrac{5}{12} f^2_{k-1} + \tfrac{3}{8} f^3_{k-1/2}. \tag{A.II.15}$$

By putting $k=2$ we immediately get

$$\Delta_{5/2} = f_2 + \tfrac{1}{2} f^1_{3/2} + \tfrac{5}{12} f^2_1 + \tfrac{3}{8} f^3_{3/2}$$

which is the third of Krylov's formulae. Now notice that

$$f_{k-1}^2 = f_k^2 - f_{k-1/2}^3 = f_k^2 - f_{k+1/2}^3 \, .$$

Substitute this value for f_{k-1}^2 in (A.II.15) and remember that, according to our assumption $f_{k-1/2}^3 = f_{k+1/2}^3$. This gives

$$\Delta_{k+1/2} = f_k + \tfrac{1}{2} f_{k-1/2}^1 + \tfrac{5}{12} f_k^2 - \tfrac{1}{24} f_{k+1/2}^3 \, . \tag{A.II.16}$$

Putting $k=1$ we get the second of Krylov's formulae.

Notice, finally, that

$$\begin{aligned} f_{k-1/2}^1 &= f_{k+1/2}^1 - f_k^2 = f_{k+1/2}^1 - (f_{k+1}^2 - f_{k+1/2}^3) \\ &= f_{k+1/2}^1 - f_{k+1}^2 + f_{k+3/2}^3 \end{aligned}$$

where use has been made of our assumption that the differences of the third order are constant. Put, moreover,

$$f_k^2 = f_{k+1}^2 - f_{k+3/2}^3 \, , \qquad f_{k+1/2}^3 = f_{k+3/2}^3 \, .$$

By substituting these values of $f_{k-1/2}^1, f_k^2$, and of $f_{k+1/2}^3$, into (A.II.16) we get, after some transformations

$$\Delta_{k+1/2} = f_k + \tfrac{1}{2} f_{k+1/2}^1 - \tfrac{1}{12} f_{k+1}^2 + \tfrac{1}{24} f_{k+3/2}^3 \tag{A.II.17}$$

which, for $k=0$, gives the first of Krylov's formulae.

Let us finally add that there are also other methods by which our problem could be solved. For a review of them the especially interested reader may consult the book

Collatz, L.: 1966, *The Numerical Treatment of Differential Equations*, Springer, Berlin.

INDEX OF NAMES

(Number in italics refers to the page on which the reference is listed.)

INDEX OF SUBJECTS

ASTROPHYSICS AND SPACE SCIENCE LIBRARY

Edited by

J. E. Blamont, R. L. F. Boyd, L. Goldberg, C. de Jager, Z. Kopal, G. H. Ludwig, R. Lüst,
B. M. McCormac, H. E. Newell, L. I. Sedov, Z. Švestka, and W. de Graaff

16. S. Fred Singer (ed.), *Manual Laboratories in Space. Second International Orbital Laboratory Symposium*. 1969, XIII + 133 pp.

17. B. M. McCormac (ed.), *Particles and Fields in the Magnetosphere. Symposium Organized by the Summer Advanced Study Institute, held at the University of California, Santa Barabara, Calif., August 4–15, 1969*. 1970, XI + 450 pp.

18. Jean-Claude Pecker, *Experimental Astronomy*. 1970, X + 105 pp.

19. V. Manno and D. E. Page (eds.), *Intercorrelated Satellite Observations related to Solar Events. Proceedings of the Third ESLAB/ESRIN Symposium held in Noordwijk, The Netherlands, September 16–19, 1969*. 1970, XVI + 627 pp.

20. L. Mansinha, D. E. Smylie and A. E. Beck, *Earthquake Displacement Fields and the Rotation of the Earth. A NATO Advanced Study Institute. Conference Organized by the Department of University of Western Ontario, London, Canada, 22 June–28 June, 1969*. 1970, XI + 308 pp.

21. Jean-Claude Pecker, *Space Observatories*. 1970, XI + 120 pp.

22. L. N. Mavridis (ed.), *Structure and Evolution of the Galaxy. Proceedings of the NATO Advanced Study Institute, held in Athens, September 8–19, 1969*. 1971, VII + 312 pp.

23. A. Muller (ed.), *The Magellanic Clouds. A European Southern Observatory presentation: principal prospects, current observational and theoretical approaches, and prospects for future research. Based on the Symposium on the Magellanic Clouds held in Santiago de Chile, March 1969, on the occasion of the Dedication of the European Southern Observatory*. 1971, XII + 189 pp.

24. B. M. McCormac (ed.), *The Radiating Atmosphere. Proceedings of a Symposium organized by the Summer Advanced Study Institute, held at Queen's University, Kingston, Ontario, August 3–14, 1970*. 1971. XI + 455 pp.

In preparation:

25. G. Fiocco (ed.), *Mesospheric Models and Related Experiments. Proceedings of the 4th ESRIN-ESLAB Symposium, held at Frascati, Italy, July 6–10, 1970*.

27. C. J. Macris (ed.) *Physics of the Solar Vorona. Proceedings of NATO Advanced Study Institute on Physics of the Solar Corona, held at Cavouri-Couliagmeni, Athens, Greece, 6–17 September 1970*.

SOLE DISTRIBUTORS FOR U.S.A. AND CANADA

SPRINGER-VERLAG NEW YORK, INC., 175 Fifth Ave., New York, N.Y. 10011